HELPING YOURSELF WITH
NEW ENZYME CATALYST
HEALTH SECRETS

Other Books by the Author

Health Tonics, Elixirs and Potions for the Look and Feel of Youth
Natural Hormones: The Secret of Youthful Health
Health Secrets from the Orient
Magic Enzymes: Key to Youth and Health
The Miracle of Organic Vitamins for Better Health
Miracle Protein: Secret of Natural Cell-Tissue Rejuvenation
All-Natural Pain Relievers
Brand Name Handbook of Protein, Calories and Carbohydrates
Slimfasting — The Quick "Pounds Off" Way to Youthful Slimness
Healing and Revitalizing Your Vital Organs
Miracle Rejuvenation Energizers
Encyclopedia of Power Foods for Health and Longer Life
New Enzyme Catalyst Diet: The Amazing Way to Quick,
  Permanent Weight Loss

HELPING YOURSELF WITH NEW ENZYME CATALYST HEALTH SECRETS

Carlson Wade

FOREWORD BY WILLIAM S. KEEZER, M.D.

Parker Publishing Company, Inc.
West Nyack, New York

Library of Congress Cataloging in Publication Data
Wade, Carlson.
  Helping yourself with new enzyme catalyst health
secrets.

  Includes index.
  1. Enzymes—Therapeutic use. 2. Food, Natural—
Therapeutic use. I. Title.
RM666.E55W33   615'.35   80-28081
ISBN 0-13-386904-0

Printed in the United States of America

Dedication

To Your New and Healthy Body

Foreword by a Doctor of Medicine

The New Enzyme Catalyst Health Secrets are a dynamic breakthrough in the search for body-mind rejuvenation at any age.

Complete in this highly recommended volume, you will discover how to use simple enzyme-powerful foods and programs to supercharge your body with amazing vitality and youthful health.

Carlson Wade, a leading medical writer and highly respected nutritionist of distinction, has prepared a most helpful book on new and all-natural methods of total healing.

Chapter after chapter, page after page, he tells you how you can free yourself from "hopeless" ailments...often you can experience refreshing revitalization in just a matter of minutes.

Carlson Wade shows how you can wake up your sluggish responses, release your built-in (but sleeping) self-rejuvenation catalysts, and emerge with vitality-filled health.

Helping Yourself with New Enzyme Catalyst Health Secrets spotlights the little-known reason why you age. Then the book shows you how to use cell-tissue revitalization to actually reverse the aging process and become reborn.

This major new advance in the search for natural holistic healing shows how you can free yourself from arthritis pain with soothing enzymes. You are told how to enjoy a "schoolperson's complexion," how to be "totally young" with enzyme-energized glands, how to energize your digestive organs with a simple enzyme-collagen program. Also, how to wash your arteries, relieve your headaches, melt away pound after pound, reduce inch after inch, improve your sight and hearing, and much, much more.

You will be able to repair and regenerate your vital body organs with the use of these easily followed and quickly effective enzyme-collagen programs.

7

Carlson Wade is to be highly commended for preparing this treasury of enzyme healers for just about every part of your body.

With the help of Carlson Wade's New Enzyme Catalyst Health Secrets, you are able to actually eat and drink your way to total youth.

This is the most useful self-help book on youth restoration I have ever seen. Consider it as having a specialist at your fingertips. Use this book daily. You will soon be able to look at yourself in the mirror and exclaim, "I never looked so young…and felt so good as I do now!"

William S. Keezer, M.D.

What This Book Will Do For You

This book will show you how to naturally correct most common and uncommon health disorders that do not properly respond to medications.

Most likely, you have gone the route of various drugs, medicines, and habit-forming chemicals but with unsatisfactory results. When their effects wear off, your symptoms return, often in more severe discomfort. So it is only natural for you to search for a health program that will give you freedom from pain without drugs.

This book offers you that program. It shows how you can "wake up" your sluggish metabolism; it shows how to send forth an enzyme catalyst reaction that soothes the cause of your ailment and thereby creates inner healing. Reward? A "forever healthy" feeling from head to toe.

If you have been troubled with "hopeless" problems that stubbornly refuse to yield to chemotherapeutic medical treatments, then you need this book. It shows you how to use all-natural methods, foods, and programs (some take a few minutes) to stimulate your catalyst responses and give you quick relief.

But even more important is that this all-natural and simple enzyme catalyst program gives you long-lasting benefits. Once your body's metabolic processes have been refreshingly rejuvenated with the use of any of these secrets, you can say "goodbye" to so-called "hopeless" ailments that you thought would never go away.

This book is unique because it tells you how to promote swift healing (often before your eyes) without the use of high risk medications. Instead, you use everyday foods, many of which are easily available in your neighborhood. You probably have most of these items in your own pantry. Their cost is small. Many secrets are absolutely free!

9

Follow them anywhere, anytime. At home, at work, at play, you can use these enzyme catalyst methods to give you "instant" relief.

This book will show you how to enjoy a renewed lifetime of youthful health and stamina with these secrets. Say "hello" to your new era of refreshing good health. It's yours for the taking!

SPECIAL BONUS: Need quick help for a problem? Just turn to the Enzyme Revitalization Index in the back of this book. Its unique "dial-a-health-remedy" arrangement will give you the suggested enzyme solution for your ailment...in seconds.

Carlson Wade

CONTENTS

*The new program introduced • How enzymes are able to
create quick revitalization of the body and mind • How
enzymes rebuild and maintain youthful health • Six
enzyme catalyst foods that revive the "Youthing"
process • From "Aging" to "Youthing" in a few days •
Total rejuvenation with enzymes, the "Take Charge" catalysts*

*Enzymes: the source of life • Where to get your
youth-restoring enzymes • Enzyme functions at a glance
• An enzyme Energy elixir • More enzymes needed in
older years • Twelve super-strong enzyme foods for
youthful regeneration • How to get top enzyme value
out of raw vegetables • Your emotions affect your
enzymes — be cheerful and relaxed when you eat*

*Activated enzymes alert your body cells to send forth
soothing, cooling, natural cortisone • The enzyme
program that creates miracle arthritis healing • Raw
milk is important • Enzymes + nutrients = arthritis
healing • Your lifetime "Freedom from Arthritis"
program • Foods to avoid • A raw food
"Enzyme-Energizing" program • How enzyme-activated
digestion soothes arthritis pain*

CHAPTER 1

Six Enzyme Catalyst Foods That
Slow Aging and Start "Youthing"

You can improve your hopes for a longer life — and add 10, 20, 30 or more health-packed, youth-filled years to your life expectancy.

A Special Bonus Awaits You. These are more than ordinary years. These are healthy, happy years that are free of discomfort and pain. These are years of happiness that accompanies an ache-free lifestyle. You can have this power-packed "youth bonus" simply from unique protein-like substances found in an assortment of everyday foods that go to work, almost instantly, to provide total rejuvenation so that you can help your body slow (even halt) aging and start "youthing." This promise of youth is just one part of the special bonus that awaits you with the use of these substances.

Vitality, Youthful Appearance, Total Health. Once these youth-creating substances start the process of cellular revitalization, they set off a series of beneficial biological chain reactions throughout your body. They help boost sluggish metabolism, improve gland-hormone function, regenerate organs of digestion, wash your arteries, melt away body fat and refresh your circulatory system so that you feel "reborn." Often, you can feel the benefits within moments. Nearly always, you will experience joyful vitality within hours. You will present a more youthful appearance as these "youth workers" help promote a chain reaction within your body (often in just thirty minutes!) to boost your look and feel of total health.

With the use of these "youth factors" you will be able to awaken each morning filled with the love of life and the energy of an athlete. You will be teeming with unbelievable vitality. It will, indeed, be a dream come true.

All this — and more — is waiting for you in a set of foods in the form of what are known as *enzymes*. They are able to add health-packed, carefree years to your life. Nearly always, within a few hours; but always, within a few days. *Enzymes* are the wonder workers of rejuvenation that offer this promise of renewed health and amazing vitality.

ENZYMES — MIRACLE YOUTH FACTORS
FOR TOTAL REJUVENATION

What Are Enzymes? Enzymes are protein-like substances found in all living substances. They are powerfully abundant in fresh raw fruits, vegetables, grains and seeds. They are also found in your body where they rule *all* processes. Enzymes are so influential and powerful that they determine your levels of health and feelings of youth.

How Enzymes Rebuild and Maintain Youthful Health

When you eat a raw food (or drink its raw juice) you send a high concentration of powerful enzymes into your digestive-metabolic system. These enzymes immediately set upon the task of digesting all foods and then transporting their nutrients (vitamins, minerals, proteins, carbohydrates, fats, etc.) to all parts of your body to bring about youthful regeneration.

Enzymes will also perform these essential life-creating actions:

1. Enzymes use digested foods to build healthy muscles, nerves, bones, and glands.
2. Enzymes transport nutrients for storage in your liver and muscles for future use.
3. Enzymes are involved in the formation of urea (end product of the decomposition of proteins) that is later to be washed out of your body through eliminative channels.
4. Enzymes help eliminate carbon dioxide from your lungs.

5. Enzymes help fix iron in your blood cells. Even if you have an abundance of iron-containing foods in your diet, this mineral may remain "dead" unless enzymes are available to metabolize and digest it and help build it into your bloodstream.
6. Enzymes attack waste materials in your blood and tissues and put them into a form that can later be eliminated.
7. Enzymes cause coagulation of your blood so that wounds are healed more rapidly.
8. Enzymes help decompose toxic hydrogen peroxide and release healthful oxygen from this poisonous waste.
9. Enzymes are needed to change protein into usable amino acids, to change carbohydrates into energy and to transform fats into sources of warmth and cell-building regeneration.
10. Enzymes hold the key to life with their powers of being able to heal and rejuvenate, aid digestive powers and promote assimilation and elimination. *Only enzymes can break down ingested foods into nutrients which can then be absorbed into your bloodstream to promote rejuvenation of every part of your body.*

WITHOUT ENZYMES THERE WOULD BE NO LIFE

If enzymes are totally absent, then life would cease. If there is a *partial* reduction in the availability of enzymes, then life is only *partially* possible. If there is a *weakness* in your body's enzyme system, then you in turn become weak in terms of external and internal health; therefore you can easily appreciate the power of enzymes. They do hold the key to your hope for better health and longer life.

With Enzymes, Total Youth Is Possible. When you boost your intake of enzymes, when you give your body the working materials required to create new cells for old, to regenerate "aged" organs, to revive sluggish processes, then you open the floodgates to an onrush of the rivers of life. With the use of active-energizing enzymes, total youth is possible. These enzymes are found in a variety of raw foods. More important is that each enzyme is *specific*. This means that certain foods have more potent youth-building powers than others. Also, certain enzyme-containing foods are specific for the regeneration of individual body parts or functions.

SIX ENZYME CATALYST FOODS THAT
REVIVE THE "YOUTHING" PROCESS

Why Do They Work? They work because enzymes are *catalysts*. That is, they perform an action, but do not themselves become part of that action. This makes them "rulers" of your health. *Example:* You light a fire to cook some food. The fire is comparable to an enzyme in that it cooks the food but does not become part of it. You don't eat the fire! Yet, without the fire, you would not be able to cook the food. This indicates the *catalyst* power of enzymes.

More Examples: You have green tomatoes that need to ripen. You put them on a windowsill where the warmth of the sun causes the enzymes to function and turn the tomatoes into a nutritious red. Also, a green banana is left at room temperature for a day or so. Natural warmth and moisture in the banana alert enzymes so they can transform the fruit into a golden edible delight. A pear is almost rock-hard when picked from the tree. Put it on a table and wait a few days, it becomes more than just soft; it becomes juicy and brimming with vitamins and minerals that were not previously available. Enzymes transformed the inedible pear into a nutritious treat.

Enzyme Foods Create Living Health...in Minutes. With the use of any or all of these six unique enzyme catalyst foods, you can spark the processes within your own digestive-assimilative system to create living health — more important, *youthful health!* Within minutes, or at the most, a few hours, you'll be able to say "goodbye" to the "aging" process and "hello" to the "youthing" response. Let's look at these foods and see how they work their miracles of enzyme catalyst regeneration.

#1. Papaya — Powerful Digestive Aid

Benefit: This gold-colored tropical fruit contains *papain,* a powerful digestive enzyme. When you eat this luscious fruit, you immediately alert your digestive glands to release more enzymes to metabolize foods that you are going to eat. The *papain* in this fruit will "attack" the strongest of ingested foods (such as meats) and break them down and remove important nutrients that can then rebuild your health.

Youthful Digestion Restored in Ten Minutes

Jane W. looked haggard and worn, with pale skin (characteristic of anemia) and was always complaining that most foods just "did not agree" with her. A typical meal gave her a "heavy" feeling in her stomach. She had poor digestion problems; malnutrition made her look and feel much older than her middle years. A nutrition-minded gastroenterologist (specialist in ailments of the stomach) suggested that she *finish* every meal with a few slices of freshly washed and sliced papaya. Jane tried this program because she always had serious aftereffects from chemicalized tablets or fizz powders. *Results?* Within ten minutes, she exclaimed that the heavy feeling from the steak, mashed potato and gravy dinner she just finished was gone. She smiled. She looked and felt better. Her skin glowed. She had "instant" energy. Jane W. now finishes every meal with papaya slices. The enzymes have revitalized her digestive process. Nutrients no longer "lie dead" in her system, but are enzyme-propelled throughout her body where they create miracle youthful reactions.

#2. Pineapple — Healer of Body Organs

Benefit: This spiked tree-borne fruit from the Caribbean and Pacific is a powerhouse of more than just vitamins and minerals. It contains *bromelain,* a unique enzyme that is able to heal, knit, bind and repair the delicate billions and trillions of body cells and tissues that make up your many body organs. The bromelain is also able to take Vitamin C and potassium, join them with Vitamin A (from the pineapple) and then use this combination to create cellular regeneration of your organs.

Stomach Upset Calmed Down in Thirty Minutes

Salesman Martin L. complained that whatever he ate "reacted" on him. He found it difficult to accommodate fat-containing foods. This indicated a weakness in his gallbladder and also his liver. Patent medicines gave him such serious "acid indigestion" that he would moan and groan with gaseous pain for hours afterwards. An herbal pharmacist suggested he try eating fresh pineapple slices as dessert after every meal. Martin L. did so. For the first time, his favorite flour-coated chicken and gravy meal did not act upon him. All he did was eat a few fresh pineapple slices after the meal and he felt that his stomach was "happy." He felt young, too, because now he could eat with the youthful digestion of an adolescent.

Benefits: The *bromelain* in the pineapple is able to help repair-regenerate-revive the functions of the fat-melting components of the gallbladder and liver. Within *minutes,* this enzyme activated sluggish metabolic processes. The bromelain enzyme took the cell-building Vitamin C, the muscle strengthening potassium and the membrane-restoring Vitamin A in a unique combination. The enzyme propelled this combo to the "needy" gallbladder and liver and revitalized them. Within moments, these two organs could effectively metabolize fat and other nutrients in ingested food. The enzyme helped in the assimilation of these nutrients so that there was a feeling of contentment and not overloading during the digestive process.

For quick revitalization-rejuvenation of body organs, the bromelain enzyme in the pineapple is a miracle of healing.

#3. Green Leaves — Source of "Forever Young" Health

For the look and feel of "forever young" health, discover the powerhouse of enzymes in a plate of green leaves. You may use any type of lettuce (iceberg, butterhead, Boston, bibb, oak leaf, red leaf, romaine, cos, chicory, curly endive, escarole, Belgian, spinach, watercress, dandelion, Chinese or celery cabbage, mustard greens, beet tops, kale, any type of cabbage, etc.). You will be transporting a supply of speedily assimilated enzymes that will perform dynamic youth restoration...often within moments.

How to Use: Before your main meal, prepare a bowl of seasonal fresh green leaves in a simple salad. For taste satisfaction, try different varieties. Tear the leaves or chop them finely. Add your favorite dressing.

How to Eat: You begin by *chewing thoroughly.* This is highly essential for enzyme effectiveness. The chewing act causes the release of salivary enzymes which help break down the cellulose fibers of the leaves and release their invigorating nutrients. At the same time, *chewing* activates your digestive glands so that a sufficient supply of hydrochloric acid (enzyme) is released and is then able to metabolize swallowed food. The enzymes released from the green leaves act as strong forces of energy to distribute metabolized nutrients to all body parts. That is why *chewing thoroughly* is essential for enzyme effectiveness.

How Green Leaves Enzymes Keep You "Forever Young." These effective enzymes are able to transport "hard to get" amino acids throughout your system and create so-called "complete protein" nourishment. Green leaves enzymes are alkaline in body reaction and are therefore soothing. These enzymes also take the chlorophyll (mineral-like coloring matter) out of the green leaves and use this to stimulate your bone marrow to produce youthful hemoglobin. Your bloodstream becomes rejuvenated in moments.

More Benefits: Green leaves enzymes are then used to help you digest and (especially important) *utilize* food. This builds your resistance to ailments and so-called aging. Green leaves enzymes will stimulate a healthy bowel action, improve your heartbeat, bring about cell-tissue regrowth (the basis of perpetual youth) and create a feeling of youthful warmth.

As an enzyme food, green leaves are high concentrates of this dynamic catalyst. A bowl of such green leaves, taken just once a day, can help firm up your skin, soothe your digestive unrest, restore youthful agility to your muscles and give you vitality and energy, often within a day or so. Enzymes work swiftly. This is the prime reason for their success in creating miracle rejuvenation... in a short time.

#4. Grapefruit — Natural Body Cleanser

The average grapefruit has a tart, almost acidic taste, which indicates its strong reaction in your body. Enzymes of the grapefruit combine with Vitamin C and then scour your trillions of cells and tissues. The strong grapefruit enzymes will wash off accumulated toxic wastes and free radicals (waste elements that cling to your vital organs and block free nutrient passage) and help remove them from your body.

Once you are internally cleansed, your grapefruit enzymes use Vitamin C to create *collagen,* a connective tissue substance that cements, binds and repairs your trillions of body cells. Within a few hours after taking grapefruit, the enzymes and Vitamin C have created this cleansing-repair process. You'll *feel* totally cleansed. Your skin will *look* healthy and sparkling fresh. You will glow with new youth.

One-Day Grapefruit Fasting Program

At age 60, Rose J. looks less than half her age. Reason? Every 10 days, she goes on a simple grapefruit fasting program. It was advised by her nutrition-minded physician when she complained of feeling tired all the time. Also, her skin started to sag. She became more vulnerable to infections, colds, respiratory distress.

Rose J. needed "internal cleansing" and "internal regeneration" in a hurry. So she followed the nutritionist-doctor's program: every 10 days, devoted just one day to eating grapefruits and drinking freshly prepared grapefruit juice. This totally revitalized Rose J. Now she had youthful energy. Her skin firmed up. She had unbelievable immunity to common and uncommon ailments. She felt she was in her second or "new" youth. All this, thanks to grapefruit enzymes.

Simple Program To Follow: You may plan a regular "grapefruit fasting day" to help reconstitute your damaged cells and protect against degeneration with the help of the fruit's enzymes and vitamins. Also, plan to eat grapefruit wedges as part of a salad, at least once a day. Enjoy grapefruit juice, too. You'll find your body being alerted by the enzymatic action that works almost immediately. You'll then be able to grow younger...in minutes.

#5. Apple — For the Glow of Youth

A freshly washed apple is the source of powerful health. Enzymes in the apple will help improve the functioning of your body's digestive and regulatory systems. Enzymes will take the pectin and hemicellulose and the minerals in the apple and use them to create a healthy acid-alkaline balance. This is the key to youthful digestion and metabolism.

Youthful Resistance To Aging Ailments. Enzymes will take the high potassium supply and use it for washing your cells and tissues; also, for the proper maintenance of mineral balance in your bloodstream. This creates a "buffer zone" so that infectious bacteria cannot penetrate and cause so-called "aging ailments." Apple enzymes create this catalytic action by fighting body toxins, boosting digestion and rejuvenating the entire body system.

Natural Cholesterol Regulator. Apple and your own digestive enzymes now combine to take the pectin from this fruit and help your body maintain a natural cholesterol level. By this self-regulation, you are able to have youthful protection against the risks of heart distress.

Creates Body Detergent Action. Enzymes in the apple create a mild fibrous reaction for body detergent action. The apple enzymes are known for being natural dentifrices and also act as a natural "toothbrush." Eat one or two apples after a meal and the enzymes will help scrub away sticky residues between and around your teeth. At the same time, digested apple enzymes help "scrub brush" away debris clinging to your body organs. This helps give you a cleansed system. You now have a feeling of glowing youth. Your entire body will respond with crisp-clean youthful health. You will, in effect, glow like a shining apple!

#6. Banana — Miracle Youth Food

All-Natural Medicine. In this golden fruit lies a treasure of enzymes that can create near-miracle powers of total healing.

Unique Power: Aging is often traced to hypoproteinemia (protein deficiency or inadequate protein metabolism due to enzyme weakness). If neglected, this protein deficiency can cause sagging skin, body organ degeneration, slow-up of most functions and lowered resistance to infection. You can protect against this aging threat by eating several bananas daily.

The protein-sparing action of the natural carbohydrates and vigorous enzymes in the banana will accelerate the use of protein in the body. The banana enzymes will then activate and enforce the *use of ingested protein* (in the form of enzyme-transformed amino acids) *as building blocks for your entire body.*

Helps Soothe Peptic Ulcer

Troubled with the burning sensation of a peptic ulcer, Ralph E. asked a registered dietician for help. He wanted to soothe ulcer flareup in a simple but quick way. Medications, Ralph E. complained, only worsened his problem. The dietician advised Ralph E. to eat three to four bananas daily. *Special Advice:* Whenever any burning sensation is felt, eat a banana. *Results?* Within three days, he found relief from the burning reaction. Now, he has bananas daily and feels as if he never had an ulcer in the first place.

Special Benefits: The banana enzymes are able to neutralize or buffer runaway hydrochloric acid that causes the burning upon the open sore that is an ulcer. Its bland consistency, soft texture and low residue make it most soothing.

Enzymes offer protection of the mucosa (stomach membranous covering) from harsh irritants and from burning gastric juice. Enzymes may then be considered as natural medicines in the banana.

FROM "AGING" TO "YOUTHING" IN A FEW DAYS

By using these enzyme foods daily, you can help halt and reverse the aging process and then start "youthing" through the power of cellular rebuilding made possible by these miracle workers.

Simple Program: Every day, enjoy as many of the preceding six enzyme catalyst foods as possible — as snacks, before a meal, with a meal or after a meal. The enzymes will be able to perform this quick regeneration in a little while. You will "bounce back" with renewed vigor and a "forever young" appearance.

7-Step Enzyme Catalyst Revitalization Program

Many health spas feature enzyme programs as part of their revitalization remedies. You can now follow a 7-step enzyme catalyst program as outlined for guests at the Meadowlark Health Center of California. According to its founder, Evarts Loomis, M.D., this 7-step program is able to create a look and feel of youth, almost from the start.[1] It is quick to produce beneficial results. It is easy to follow:

1. Eat a large amount of raw fruits and vegetables daily.
2. Begin each meal with a raw food such as a salad, a fresh fruit or vegetable, or a glass of fresh juice. The raw food enzymes help you digest your meals more easily.
3. Eat raw seeds and nuts daily. Add brewer's yeast and wheat germ to salads.
4. Avoid processed, packaged or canned foods.
5. If vegetables must be cooked, steam them for a very short time to prevent excessive loss of enzymes and other nutrients.
6. Avoid fried foods, preservatives and artificial sweeteners.
7. Include whole-grain breads and cereals in your daily food program.

Youth Suggestion: Include this 7-step enzyme catalyst revitalization program in your daily food program and you will help *build and re-build* your powers of self-rejuvenation.

FOR TOTAL REJUVENATION, TRY ENZYMES

Enzymes are the "take charge" catalysts. They are "activists" and youth builders. They are absolutely essential to youthful life. They are the life of any food. They are colloidal combinations with the cell proteins and create tissue catalyst reaction — the key to total rejuvenation.

---

[1]*Natural Health Bulletin.* Parker Publishing Co., West Nyack, New York 10994. Vol. 8, No. 20. September 25, 1978. Available by subscription.

Miracle Youth Workers. These enzymes are present in all living things. All activity of life depends upon them. The greening of the leaves in the spring, the coloring of leaves in the fall, the ripening of foods, all require enzymes.

Without enzymes, seeds would not sprout, the good earth would not produce, fruits and crops would not ripen, let alone grow! So it is with your body. Without enzymes, you could not survive!

All Enzymes Needed. You need *all* enzymes to maintain healthful rejuvenation. A missing enzyme can cause susceptibility to illness and aging. There is a specific gene, an inheritance factor in the chromosome to govern production of each enzyme in your body. When the enzyme is missing or defective, the genetic makeup falters. This paves the way to deterioration.

Gradual, Slow Aging. A deficiency or weakness of any enzyme does not manifest itself overnight. Instead, it is the day-to-day decline in health, the gradual weakness that increases with the passage of time, the heightened susceptibility to illness, the erosion of body processes. It is this gradual, slow aging that is caused by enzyme deficiency.

Reverse This Aging Tide. Bolster your enzyme intake through fresh raw fruits, vegetables, grains, seeds and nuts. In so doing, you will be fortifying and strengthening your trillions of body cells and resisting the unnecessary approach of so-called old age.

With the use of these enzymes, you will no longer look and feel "half alive." Instead, you will be suddenly flooded with healthy cheer and an urge to enjoy life to the fullest, all because of the power of enzyme catalyst foods that are able to help you slow aging and start "youthing"!

HIGHLIGHTS

1. Lengthen your life with youth-filled years through the help of enzyme catalyst foods.

2. Note the 10 unique benefits available with the use of enzymes.

3. There are six special enzyme-packed foods that offer hope for healing and rejuvenation.

4. Jane W. enjoyed restoration of youthful digestion, firmed-up skin and better health, in minutes, with the use of papaya.

5. Martin L. calmed down his perpetually upset stomach and irritated gallbladder and liver with the use of pineapple enzymes.

6. Rose J. follows an occasional one-day grapefruit fasting program. She looks half her age because grapefruit enzymes cause rejuvenation of trillions of body cells.

7. Ralph E. feels that his peptic ulcer no longer exists because of the soothing action of tasty banana enzymes.

8. Follow the 7-step enzyme catalyst revitalization program, just as performed in a famed health spa, and become rejuvenated in days.

CHAPTER 2

A Dozen Enzyme-Strong Foods That Energize Your Body and Mind

Increase the *quantity* and *quality* of enzymes in your digestive system, and you set off a series of beneficial reactions within your entire body so that you soon feel body-mind rejuvenation.

Enzymes: Source of Life. These miracle youth workers make your life possible. They are required for every biological cell-organ reaction that occurs in your body. Vitamins, minerals, proteins and hormones *must* have enzymes to perform their healing processes.

Unique Rejuvenating Power. Enzymes are likened unto a "building force" that shapes your body just as construction workers are a force that builds your house. *Note:* You may have all the required building equipment, but to put it into shape, you need workers. So it is with your body. You may eat nourishing food that is brimming with nutrients but to put them into all your cells and systems, you need workers: namely, enzymes. *Your entire life depends upon these enzymes.*

WHERE TO GET YOUR YOUTH-RESTORING ENZYMES

Basic Source: Fresh, *raw,* uncooked fruits, vegetables, grains, seeds, nuts — and freshly prepared juices from these sources — are excellent suppliers of these building forces.

Caution: Cooking Will Destroy Enzymes. Although powerful, enzymes can be weakened and destroyed by heat. Temperatures above 118°F. will inactivate and destroy these youth-restoring enzymes. Any product that has been pre-heated, pre-cooked or subjected to these high temperatures will *not* contain enzymes. You should emphasize a selection of fresh and *raw* foods in order to supply your body with these essential life-giving substances. Balance it with cooked foods, of course, but your goal should be a wholesome quantity of raw foods every single day. Since enzyme rejuvenation is an ongoing process, you should give your body a never-ending supply of these youth workers with daily intake of tasty *raw* foods.

Enzymes Have Built-in Eternal Life

Enzymes, in their dormant state, as found in dry seeds, seem to have a natural built-in eternal lifespan. Enzymes have been found in rice stored for hundreds of years. Researchers have found living enzymes in Egyptian tombs over 3000 years old. Enzymes have also been found in the frozen flesh of ancient mammoths, buried in solid blocks of ice and snow in the regions of Siberia. This would make the enzymes at least 50,000 years old! While enzymes can survive and thrive at sub-zero temperatures, they are completely destroyed at temperatures over 118°F. So in planning your enzyme rejuvenation program, emphasize raw and uncooked foods. These are prime sources of enzymes that appear to have a built-in eternal life!

Enzyme Deficiency Can Cause Premature Aging

If you have a low supply of these miracle workers, then you force your body to draw upon its enzyme reserve. This causes a strain on your enzyme "bank" and is the underlying cause of many so-called degenerative diseases. This deficiency causes a body breakdown and the onset of undesirable (and unnecessary) premature aging.

Deficiency Causes Body Strain. If your body needs to create more enzymes from storage depots, this causes a strain. This "stealing" of enzymes from different areas of your body to be put into your digestive tract can cause internal "competition" for these precious youth workers, and your different organs, systems and cells start to "battle" for their existence. This can cause eventual degeneration and strain and the ravages of premature aging.

Illness May Result from Enzyme Confusion. Because of this metabolic upheaval, your body systems can no longer function as healthfully as they should. An enzyme shortage occurs. There is rampant confusion amongst your various internal organs and systems. This may cause a *biological upset* which is the forerunner of many so-called illnesses of civilization ranging from heart upset to cellular destruction. Therefore, you need to keep your "bank" supplied with quantities of adequate enzymes so that you maintain body equilibrium and harmony.

ENZYME FUNCTIONS — AT A GLANCE

Digestive Power. Enzymes are small biochemical digestants. They have the individual power to cut apart vitamins, minerals, proteins, carbohydrates, fats, etc., and make them absorbable. Remember that enzymes do not become part of the metabolized end product but cause it to be manufactured. Therefore, you receive *pure* digestible nutrients made possible solely by your enzyme quantity and quality.

Recognizing Enzymes. You have some 8000 different enzymes in your body. You can recognize them because nearly all are chemically or biologically named so that they end in *-ase*. For example: protein-splitting enzymes are *proteases*. Fat-splitting enzymes are *lipases*. Carbohydrate-splitting enzymes are *amylases*. Generally speaking, this is the easy way to recognize enzymes — at a glance.

Full Supply Is Needed — Substitutions Are Inadequate

Each enzyme has its individual or specific function. A full supply is needed. Substitutions are inadequate. For instance, *lipases* are needed to split and help digest ingested fat. If you're weak in this enzyme group, the other groups will not be able to do this fat-splitting task and you run the risk of excessive deposits in your system. So you need a full daily supply.

Some enzymes *build up* new materials in your body to create the ongoing regenerative process. Other enzymes are used to *break down* wastes and help to eliminate them from your body as a scrubbing action. You need *both* classes of these enzymes else there is a disturbance and imbalance and the regenerative-rejuvenation process becomes inadequate.

THE ENZYME THAT GIVES YOU YOUTHFUL ENERGY

To enjoy youthful body-mind energy, you need one enzyme group called *transketolase,* found in most raw grains and also fresh raw vegetables and their juices. If you want speedy revitalization, you can easily prepare a simple tonic right in your own home that is every bit as tasty as it is energizing.

Enzyme Energy Elixir

Combine fresh raw celery, carrots, cabbage with two tablespoons of wheat germ or whole bran (or combo.) Blenderize for one minute. Drink slowly. Within 30 minutes, you will find yourself looking and feeling energized in body and mind.

Becomes Young Again

Schoolteacher Vera T. was often faced with midmorning (and midafternoon) fatigue. Classwork could be tiring. She complained to the school nurse who recommended this simple tonic. Vera T. prepared it in the morning and put it in a thermos. During morning and afternoon recess, Vera T. took just one glass of this *Energy Elixir Tonic.* She was amazed at how it helped "wash away" tiredness and helped "refresh" her mind, often within 15 minutes. She returned to her class just brimming with youthful vitality. Now, Vera T. takes this tonic twice daily and has all the vim and vigor of a teacher much younger than her 50-plus years. She gives it a good mark on her personal health report card!

Benefits: The *transketolase* enzymes in the juice energize the B-complex vitamin, thiamine (B1) in the grains and release supplies of magnesium to break down dietary carbohydrates and send forth a superior source of energy, almost instantly. *Bonus:* Prepared in juice form, the *transketolase* enzymes are able to work more speedily and become assimilated almost at once. This is the secret of the body-mind energizing power of the *Energy Elixir Tonic* — it carries a triple E rating!

MORE ENZYMES NEEDED IN OLDER YEARS

The so-called aging process may often be traced to the result of the slowing down and disorganization of enzyme activity. As folks grow older, their bodies tend to produce fewer enzymes. Therefore, with the increased intake of enzyme-containing raw foods and juices, you will help protect yourself against the consequences of premature aging.

Enzyme Intake More Important Than Food Intake. Many older folks make wise improvements in food intake, emphasizing more wholesome and natural products. This is helpful. But it would be more beneficial to include an assortment of enzyme-containing foods in order to make regular foods all the more nourishing.

Enzyme Power. The basic power of enzymes is in the *digestion* and *assimilation* of ingested foods. Enzymes help your body break down nourishing food into biological elements you need to maintain good health. Enzymes must always be present to digest the eaten nutrients and then transport them to build and rebuild your body to a more youthful stature.

Eat and Starve Without Enzymes. Nearly all foods you eat are totally indigestible until enzymes work on them and break them down into simpler substances which can be absorbed in your bloodstream for eventual regeneration. *Caution:* You can eat the most wholesome and nourishing food that exists and still suffer malnutrition if you are enzyme deficient! Therefore, you need *enzymes first* and *good foods* second.

Example: To digest a cut of steak in the laboratory, to break it down into component parts that can be utilized by the body, it has to be boiled in concentrated acid for over 24 hours; then, the food value is either diminished or destroyed by overcooking. In your body, enzymes can accomplish this breaking down action within 3 hours and the end product is nourishing food! So you can readily appreciate the power of enzymes for life...knowing there is no substitute!

Simple Enzyme Test. Chew a piece of bread for a while. You will notice, gradually, a sweet taste. This is enzyme action. Your body cannot utilize starch as such; it has to be broken down to sugars. So *ptyalin,* a salivary enzyme, accomplishes this action. This same process occurs throughout your body. Your digestive enzymes "chew" food and prepare it for nourishment of your entire body.

Chew First, Digest Better. All food should be chewed thoroughly. Mouth enzymes will help break down this food so that digestive enzymes will be more effectively able to further metabolize the food for nutritional assimilation via your bloodstream. If you *chew thoroughly,* you will digest your food more enzyme-effectively. You will then be rewarded with a feeling of total energy of both body and mind.

TWELVE SUPER-STRONG ENZYME FOODS
FOR YOUTHFUL REGENERATION

Here is a set of 12 of the most important super-strong enzyme foods that offer a powerhouse of energy to help revitalize-regenerate your body and mind. They are easily available at your local market. You probably have many of them in your home right now. To begin with, follow these simple guidelines:

(1) Eat them raw. (2) Drink their freshly squeezed juices, if desired; save the pulp for a salad, too. (3) Plan to eat several of these super-strong enzyme foods every day. (4) Eat them singly or in any desired combination that is tastefully pleasing. (5) Try to eat them *before* your meals. You'll discover your food digesting more effectively. This provides better nutrient assimilation.

#1. ASPARAGUS. Contains an enzymatic alkaloid called *asparagine,* along with important minerals. *Energy Benefit:* The asparagine is able to break down Vitamins A and C in other ingested foods and use these to help improve the health of your skin. *Suggestion:* Plan to combine with carrot juice as a total tonic.

#2. AVOCADO. A prime source of most basic vitamins as well as minerals and plant proteins. *Energy Benefit:* Its natural enzymes will use the minerals and proteins to regenerate your organs and connective tissues so that you will "look and feel alive" with invigorated vitality.

#3. BROCCOLI. A prime source of calcium and phosphorus as well as potassium. Plan to eat the "heads" either chopped in a salad or steamed very briefly to make it more palatable. *Energy Benefit:* Enzymes will use these potent minerals to revive and regenerate your bloodstream and also to strengthen your skeletal structure. You'll respond with renewed vigor and clearer thinking, too, since the enzymes will use these minerals for nourishment of your brain cells.

#4. BRUSSELS SPROUTS. A high concentration of calcium as well as Vitamin C along with smaller amounts of minerals. *Energy Benefit:* Enzymes take the calcium to strengthen your nervous system so that you are able to meet the responsibilities of daily living with more youthful energy. The enzymes also use Vitamin C to knit and bind your frazzled and damaged cells and make them like new again. You'll discover a new feeling of sparkling vitality and energy in a little while.

#5. CABBAGE. Very high in Vitamin A and C as well as important minerals. *Energy Benefit:* Enzymes will use the vitamins and the minerals for the rebuilding of body cells, but particularly the digestive system and body organs. A glass or two of freshly squeezed cabbage juice will help soothe fluttery digestive action, and promote a smooth-coating action upon the gastrointestinal tract to create a feeling of "forever young" metabolism.

#6. CARROT. A golden fruit with a gold mine of nutrition. Contains carotene as well as Vitamin C and many essential minerals. *Energy Benefit:* Enzymes will use carotene to become transformed into skin-feeding Vitamin A as well as joining with Vitamin C to rebuild your epithelial tissues (cells that make up your skin, mucous membranes lining your body organs and glands) and energize these segments so that you feel vitality coursing through your bloodstream. The enzymes will also use carrot minerals for energizing your glandular organs so you enjoy healthier and more youthful hormone balance.

#7. CELERY. Good source of power-packed minerals such as copper, magnesium, manganese and phosphorus as well as Vitamins A and C. *Energy Benefit:* Enzymes will take these minerals and use them to assist in the proper functioning and strengthening of your nerves. The enzymes use the nutrients to nourish nerve pathways leading to and comprising your brain. This helps give you a youthful emotional outlook, coupled with clearer thinking.

#8. CUCUMBER. A prime source of natural silicon and also manganese together with smaller but concentrated portions of magnesium. *Energy Benefit:* Enzymes will take these minerals and bring about improved carbohydrate metabolism. In so doing, natural sugars are released into your bloodstream to provide you with almost immediate energy. You will find yourself feeling more natural vim and vigor, often within minutes after eating a cucumber.

#9. DATES. If available, use the pitted variety for more convenience. If not, carefully remove pits. A high concentration of various minerals as well as plant-metabolized glucose which is readily available for body energy utilization. *Energy Benefit:* Enzymes will take the "date sugar" and provide almost immediate availability for assimilation within your bloodstream. Eating a handful of pitted dates will provide you with youthful vitality as enzymes are able to metabolize the pre-digested glucose (tree-ripening under the sun brings about this action) almost at once.

#10. FIGS. Contain natural carbohydrates, potassium, smaller amounts of protein. *Energy Benefit:* The simple sugars in the figs (monosaccharides) in the form of dextrose and fructose are "instantly" seized by enzymes and sent directly into your bloodstream. Since natural fig sugars require almost no digestion, they can be utilized by enzymes as quickly as this super-strong enzyme food is swallowed. If you need a quick lift, if you want to act with youthful vitality, treat your enzymes to several figs. They'll provide you with quick energy, even before you finish eating them.

#11. ORANGE. A golden fruit that is a high concentration of natural fruit sugars but also skin-cell regenerating Vitamins A and C. *Energy Benefit:* Either the pulp or its freshly squeezed juice offers a treasure of enzymes that are speedily absorbed into your bloodstream. These enzymes will also use the important vitamins to nourish your skin cells and to build collagen, the cement-like connective tissue that knits and binds and holds your body together. Enzymes in the orange and its juice work speedily to create this vitality response.

#12. RAISINS. A high concentration of natural fruit sugars as well as minerals and Vitamins A and C. *Energy Benefit:* Power-packed with enzymes, this super food is able to send a supply of immediately assimilated natural fruit sugars right into your bloodstream. You'll find tiredness being washed away. You'll experience a sense of overall exhilaration, often in a matter of moments.

From "Always Tired" to "Super-Active" in Two Days

As district supervisor for a large computer corporation, Roger B. was always on the road. He loved his work, but complained of feeling weary all the time. Even at noon, when his day had hardly begun, Roger B. was starting to yawn and had difficulty in holding his eyes open. He explained his problem to a company nutritionist who suggested a simple enzyme-energizing program. Roger B. followed it for just two days. He discovered that now he had such limitless vitality, he could work into the late hours and feel as energetic as when he first started. Roger B. uses this simple but remarkably effective program daily and has gone from "always tired" to "super-active" all week long.

Easy Enzyme-Energizing Program. Carry a small packet of pitted dates, figs, raisins. Eat a handful every hour or so, or whenever you have the need for a mind-body pickup. You'll discover new energy just streaming through your mind-body connection...within moments. Enzymes will instantly deposit energy-boosting natural fruit sugars into your bloodstream which then create a total feeling of youthful energy. It's the tasty enzyme way to enjoy extended amounts of vitality.

Suggestion: To help improve your basic foundation of day-long youthful vitality, plan to eat a variety of the preceding enzyme-strong foods every single day. You will help awaken your sluggish responses and revitalize your metabolism. You will also supercharge your entire body with a feeling of exhilaration and youthful vim. You'll then discover the power that enzymes have in making you feel glad to be alive!

HOW TO GET TOP ENZYME VALUE OUT OF RAW VEGETABLES[1]

You can get extra enzyme action from raw vegetables with these suggestions. Furthermore, fresh vegetables are a marvelous source of skin-building Vitamin A, cell-building Vitamin C, nerve-nourishing calcium, blood-building iron and important roughage to maintain internal cleanliness. *Bonus:* Raw vegetables all contain enzymes which *activate* these nutrients and prod them into performing these health-youth and vitality-creating responses.

But take care. These are fragile nutrients. You want to keep them from evaporating. Here are some suggestions for doing just that:

* Use the leafy parts of some vegetables like collard, kale and turnips because they contain high concentrations of enzymes and Vitamins A and C.
* Use the outer coarser leaves of all types of lettuce, because they have higher enzyme, calcium, iron and vitamin values.

* The leaves *and* the core of a cabbage offer a high enzyme and Vitamin C value, so use *both*.
* Use a sharp blade to trim, cut or shred fresh vegetables to prevent the bruising that causes enzyme and nutrient loss.
* Refrigerate most vegetables promptly to avoid nutrient evaporation. Store them at cold-cold temperatures in your vegetable crisper or in moisture-proof bags.
* Ripen tomatoes out of the sun at temperatures ranging between 60°F. and 75°F. Ripening them in the refrigerator or on a too-hot windowsill may cause them to soften and decay and weaken their enzyme effectiveness.
* Root vegetables like carrots and sweet potatoes will retain their nutrient value if kept cool and moist enough to prevent their withering.

Should You Cook Vegetables? The answer is "yes" if they cannot be eaten raw. Obviously, a potato, pumpkin, beet, or yam, etc., is unpalatable and undesirable in its raw state. Therefore, plan to *steam* your vegetables but ONLY until crisp. Use just enough water to release steam without touching the vegetables. Be sure that the pot has a tight fitting lid heavy enough to prevent the escape of steam.

Buying, Eating, Storing. Buy only as much as you need. Plan to eat raw or steamed or cooked vegetables as soon as possible. Storing them for several days will cause some nutrient loss and defeat the purpose of having bought them fresh in the first place.

One-Day Body-Mind Enzyme Rejuvenation Program

May M. looked old beyond her years. She walked with a stooped gait. Deep crease lines gave her premature aging wrinkles. She breathed heavily upon the slightest exertion. She would feel muscular strain when doing ordinary housework. May M. thought she was just "getting old" and resigned herself to this occurrence. But a nutrition-minded neighbor told her of a dynamic, near-miracle, one-day program she followed at a well known health spa. Just one day on a total raw vegetable (solids and juices) program so revitalized her body and her mind, she looked and felt like an adolescent. She suggested May M. follow the same program in the privacy of her own home.

Bounces Back with Enzyme-Energized Vigor in One Day

May M. devoted one day to this unique food plan. It is simple. Eat no other foods. Drink no other beverages (except water and herbal tea.) Just eat fresh raw vegetables singly, in any favorite salad or in any desired combinations. Drink any raw vegetable juices as desired, either singly or in combination. No limits on amounts to be consumed. *Results?* By nightfall, May M. was walking with a springy, youthful step; her face glowed with a peaches-and-cream look. She breathed more healthfully. She could do her housework without undue exhaustion.

Just one day on the program and May M. regained her "second youth."

Suggestion: Every ten days, go on this enzyme rejuvenating raw vegetable program. It works because your digestive system is not overloaded with other foods. There is no "competition" and now enzymes can work solely upon raw plant foods to release nutrients that help regenerate your entire body and mind. It's the natural way to tap your built-in powers of self-rejuvenation.

For Super-Enzyme Power — Be of Good Cheer

Enzymes work best in a natural digestive fluid environment. This is influenced by your emotions. If you feel fearful, angered, stressful, worried, or if you are the victim of any strong emotions (or even too fatigued), then your enzymes will not work as efficiently as they should.

Emotions And Stomach Juices. Bile and hydrochloric acid are influenced by your emotions. These "stomach juices" facilitate the energy producing work of enzymes. If you're wrought-up, then there is a disturbance in the release and balance of these digestive juices. Consequently, enzyme effectiveness is weakened. *Examples:*

1. *Hydrochloric Acid.* Enzymes combine with this digestive juice to maintain the acid environment of your stomach contents. Now, water-soluble substances and proteins may be metabolized and then used to build and rebuild your body.

2. *Bile.* Enzymes combine with this liver-manufactured juice so that fats can be dissolved or emulsified. Enzymes need to split fat and can do so with the help of bile.

Problem: You eat and/or drink when you are all wrought-up. You may be physically and/or emotionally tense and you "tighten up" your digestive region. Your hydrochloric acid and bile secretions become inadequate or excessive, depending upon your tense situation. As a result, enzymes cannot work efficiently and the food you eat, no matter how top quality, will not be properly digested.

This vasoconstriction (tightening of blood vessels) can cause such enzyme distortion that your body health starts to decline. This gives rise to the premature aging and loss of energy that accompany such fatigue. But you can correct this yourself.

Be Joyful in Eating. These four words hold the key to maximum enzyme energizing of your entire body. *Be joyful in eating.* You should be happy, contented, cheerful and free of stress and strain. If not, *don't eat!* Lie down and rest for a while until you feel relaxed and then start to eat. In so doing, you will be creating a healthful hydrochloric acid and bile environment in which enzymes can perform their dynamic health-regenerating functions.

Yes, by using fresh raw foods, you can help introduce a powerful supply of body-mind energizing enzymes. They work effectively, often within hours, in helping to wash away "aging years" and revealing the "new youth" that lies beneath the shed tiredness.

MAIN POINTS:

1. Enzymes create powers of self-rejuvenation, often in a day or so.
2. Fresh raw foods are prime sources of life-giving, energy-producing enzymes.
3. Vera T. became "young again" with the use of the *Enzyme Energy Elixir.* Easily prepared, it has near "dynamite" action in boosting speedy vigor of body and mind.
4. Enjoy as many as possible of the 12 listed super-strong enzyme foods (and juices) daily for overall regeneration.
5. Roger B. ended his "tired feeling" and released his "super-active" energies with a tasty enzyme food...within two days.
6. May M. turned back the years and became young again on a one-day body-mind enzyme rejuvenation program. It is easily followed in your own home. It works within hours.

CHAPTER 3

Free Yourself from Arthritis Pain

With Soothing Enzymes

Arthritis is an error in your biological metabolism. Correct that error with the use of revitalizing enzymes, and you will be able to free yourself from arthritis pain and discomfort.

How Enzymes Offer Arthritis Relief. When you energize your weak enzyme system, you help bring about a more balanced biological metabolic response. These activated enzymes, introduced through a simple but effective raw food and supplement program, now alert the neurotransmitters located in your body cells to send forth the needed amount of natural cortisone, a hormone that is produced in the cortex (outer shell) of your adrenal glands (pair of small organs at the back of your abdomen, against the upper ends of your kidneys). The enzymes take up your body's nutrients and use them to stimulate your sluggish adrenals so that sufficient amounts of cortisone are released.

Enzyme-Activated Cortisone: Soothing, Comforting, Cooling. The enzymes stimulate the cortisone within your body to help soothe inflamed joints, comfort hurt and cool down inflammation. The enzymes are able to use cortisone to strengthen bone and fibrous connective tissue and thereby help bring about relief and freedom from arthritis pain.

Enzymes: Key To Arthritis Healing. With the use of a simple raw food program, you can help *correct* the biological metabolic error in your system that is the root cause of your arthritis distress. These enzymes will take vitamins and minerals and use them to help heal injured cells and also help bind together the osteoporosis (fragile bones condition) of the body skeleton and thereby strengthen the entire organism. When you correct the *cause* of the arthritis, you help bring about more lasting healing of the *symptoms.* Enzymes are able to strike at the cause and thereby offer you hope for freedom from arthritis.

THE ENZYME PROGRAM THAT CREATES
MIRACLE ARTHRITIS HEALING

An enzyme-aware physician, Robert Bingham, M.D., is able to heal "hopeless" arthritics in his clinic in Desert Hot Springs, California. The doctor recognizes that a deficiency of any enzyme or nutrient is a disturbance of "Nature's balance" and this causes the error that erupts as inflamed or swollen arthritis. His reported healing program is easy to follow and offers miracle healing of so-called hopeless cases[1].

Simple, Effective Arthritis-Healing Program

1. Bed rest for sixteen hours a day.
2. Increase water consumption to eight or more glasses daily.
3. Slowly reduce drug intake (doctor supervised) to levels that will cause no pain.
4. A high protein diet.
5. All foods possible are to be eaten *fresh, raw and in their natural state.* Use a blender or grinder, if necessary. *Benefit:* This supercharges the sluggish system with powerful enzymes that are able to use nutrients for healing of arthritis.
6. Eliminate all tobacco, alcohol, refined carbohydrates and saturated fats.
7. Take prescribed supplements of vitamins, minerals and enzymes.

Raw Milk Important. Dr. Bingham says that most arthritics are deficient in protein and also calcium. He prescribes three glasses daily of raw milk. This offers a good supply of these two nutrients together with enzymes that use them for correction of the metabolic error. *Note:* Dr. Bingham says that if raw milk is unavailable, use pasteurized. He prefers raw milk because it contains an *enzymatic anti-stiffness factor* that is destroyed by pasteurization.

---

[1]*Parker Natural Health Bulletin,* West Nyack, New York 10994. Vol. 10, No. 4, February 25, 1980. (Condensed version.)

Vitamin C Recommended. He prescribes 2000 milligrams of natural Vitamin C daily.

Enzyme-Containing Foods Are Powerhouses of Healing

Dr. Bingham reports that enzymes in fresh foods are the powerhouses that set off a biological chain reaction to induce healing.

It is important to preserve the natural food intact, including proteins and amino acids which have not been damaged by heat; natural hormones and enzymes which have not been altered by cooking, drying, storage or preservation, and vitamins in the highest biological efficiency. He says —

"The foods must be as fresh and ripe as possible for powerful enzyme content, grown by organic methods, free of residues of poisonous pesticides and fertilizers and delivered and prepared in as natural and palatable a form as possible."

Basic Body Improvement Programs. Dr. Bingham believes in a more holistic or "total health" approach in using enzymes for correction of metabolic errors. He feels that enzymes can promote more effective healing with these programs in daily living. Lots of fresh fruits and vegetables eaten every single day. Raw or pasteurized milk, butter, eggs, fish, cheese, meats, nuts, whole grains and seeds. Lots of bed rest. Comfortably warm mineral baths are also soothing.

Relax and Give Enzymes a Chance to Work. A fully relaxed body will enable enzymes to work more effectively. Dr. Bingham notes that many arthritics are also tense and this means their enzymes are unable to perform in a calm and orderly manner. This creates the biological error involved in metabolic defects of arthritis.

Symptoms Noted: "Nutritionists know that nervous and mental and emotional stress interfere with appetite, digestion, enzymatic absorption of food, choices of food, nutritional habits and dietary patterns," says Dr. Bingham.

"Emotional stress exerts a control on the glands of internal secretion, particularly the thyroid, adrenals and pituitary and they play an important role in bone and joint metabolism."

"Naturally, *any disturbance in the nervous and emotional function of a person* produces biological changes which can affect the bones or joints." It is this enzymatic "congestion" brought about by nervous unrest that paves the way for arthritic development. With reduced adrenal release of cortisone, the joints begin to "cry out" with pain, and inflammation and swelling starts to occur. This is the consequence of nervous distress.

Suggestion: Follow the preceding program and your nervous system will soon be so calm, you'll begin to feel soothed all over. The adrenal and other glands will then be able to release an adequate amount of comforting hormones that will pamper and relax your skeletal structure.

Simple Enzymatic Program Creates Arthritis Healing

Susan K. was examined by Dr. Bingham who says that her hands told the story. The joints of her fingers were so swollen, she could hardly work as a secretary without enduring the agonizing pain of rheumatoid arthritis. She was typical of many patients. She had no place else to go. Her disease was getting worse, not better. And no one offered the slightest bit of hope to her.

Recovers Within Six Weeks. She was put on the aforedescribed supervised program. Within six weeks, the swelling in her hands had gone down.

Healed Within Three Months. By the end of three months, *there was no trace of her arthritis.*

The healing could be traced to the copious intake of fresh raw foods and adjustments in daily life style that energized and revitalized Susan K.'s body with powerful enzymes. Now the metabolic defect could be corrected and this formerly "hopeless" arthritic was now rejuvenated and healed.

ENZYMES + NUTRIENTS = ARTHRITIS HEALING

Dr. Bingham feels that if every person were to eat three 'square meals' a day, a balanced diet, and these meals were selected for their full quota of vitamins, minerals, enzymes, protein, carbohydrates and fats in the *proper proportions,* then a person would have no nutritional deficiency.

These foods, to be fully nutritious, would have to be fresh from the farm, picked when ripe, and not preserved, canned, dried, processed, adulterated, or treated with artificial preservatives, flavorings, colorings or additives.

Dr. Bingham then asks, "But what person, in today's modern life, can eat this way every day?" This is one reason why metabolic errors occur and nutritional deficiencies lead to arthritis. He lists these nutrients which help strengthen the body to resist onset of arthritis.

Basic Program: Each and every meal must have a raw fruit or vegetable to supply enzymes. These enzymes are then able to activate the nutrients listed to help correct any deficiency. These nutrients are available in supplement form as well as in wholesome foods. Plan to combine the enzyme (raw foods) with the supplement (or food) every single day for anticipated improvement.

Suggestion: Check your health problem in the list prepared for patients by Dr. Bingham, and then note the enzyme-nutrition healer:

Osteoporosis. You need an increase of enzymes together with protein and calcium.

Osteoarthritis. Enzymes plus Vitamin D are needed.

Rheumatoid Arthritis. Enzymes plus Vitamin C are needed to correct your susceptibility to infection and inflammation.

Hypertrophic Arthritis. Enzymes plus Vitamins B-complex and E are needed to create more flexibility of your arteries and correct sclerosis (hardening) or poor joint circulation.

Degenerative Arthritis. Enzymes together with Vitamin B_6 are important to help ease distress of painful joints. This combo will help nourish your nervous system to ease the neuritis-like pain which causes vaso-constriction and interferes with joint circulation and contributes to arthritis formation.

Traumatic Arthritis. This type of pain brought about after an injury can be eased with a combination of enzymes plus zinc.

Treat Your Whole Body for Total Arthritis Freedom

"It is not just the simple deficiencies caused by a lack of one vitamin or enzyme that is important in arthritis prevention and treatment," says Dr. Robert Bingham, "but the interrelationship of ALL vitamins, minerals, enzymes and essential amino acids and fatty acids that are important. *Nature's balance* must be maintained for general health and in the treatment of arthritis."

Therefore, follow this program as part of your daily lifestyle. Build and rebuild your enzymatic system so that your powers of cleansing and resistance will be able to shield you from various illnesses, especially arthritis.

YOUR LIFETIME "FREEDOM FROM ARTHRITIS" ENZYME PROGRAM

A Toronto, Ontario, Canada physician uses a combination of cooked and raw foods as a means of helping to heal arthritis. When you follow this program, you will help supercharge your body with nutrients and enzymes that will help cleanse away ache-causing debris and help bring down swelling of painful joints. In effect, it is a lifetime program to do more than free you from arthritis. It keeps you forever free from this debilitating ailment. Leo V. Roy, M.D., of Toronto[2], offers this program to arthritis-troubled folks who seek lasting relief.

Proteins. Every meal should contain a protein. One-fourth of your daily food intake should be protein. Best sources are fish, eggs, nuts, cheese and raw certified milk.

Meats. Use internal organs, rich in vitamins and minerals. Avoid all canned and processed meat foods. Fowl, lamb and steak are good.

Dairy Products. All kinds of cheese, natural and fermented, yogurt, natural buttermilk.

Raw Vegetables. All, especially celery, cucumbers and carrots (high in enzymes). Make salads using oils for dressing.

Cooked Vegetables. Baked potatoes and brown rice are nutritious. Use raw bean sprouts and other sprouts. *Do NOT overcook any vegetables as this reduces enzyme and nutrient content.* Steam, bake or broil; use as little water as possible.

Fruits. Especially high-enzyme apples, grapes, bananas or local varieties.

Juices. Fresh-squeezed grape juice — no sugar added. No tomatoes. No citrus fruits except for flavoring. A maximum of one teaspoon lemon juice for salad dressing.

Cereals. Fresh ground combinations — wheat, rye, sesame seed, flax and millet. Do not boil. Soak fifteen to twenty minutes in hot water over a double boiler, or soak overnight and then warm up. Raw sunflower seeds, raisins or shredded coconut may be added before eating.

---

[2]Leo V. Roy, M.D. Toronto, Ontario, Canada. Press release. January, 1980.

Bread. Use sparingly. Only stone-ground fresh whole wheat or rye.

Soups. Bouillon or consomme.

Acid Drink. Where there is insufficient acid or where there are calcium deposits (such as in bursitis), use one tablespoon apple cider vinegar to a glass of water (with or without one teaspoon of honey) at least twice daily.

Sweets. Old-fashioned blackstrap molasses (unsulphured) from health food stores. Only one sweet daily, in small quantity.

Oils. Especially sesame, safflower and sunflower oil. All seeds and raw nuts.

AVOID ALL OF THE FOLLOWING

Tea	White flour
Coffee	White sugar
Alcohol	All hydrogenated (hardened) fats
Canned foods	Roasted nuts
Commercial cereals	Stale nuts
Processed foods	Stale wheat germ
Canned meats	Stale wheat germ oil

DO NOT USE

Overcooked and reheated foods.

Jams, jellies, syrups, ice cream, soft drinks, tobacco.

Chemicals added to food — sweeteners, emulsifiers, thickeners, fluoridated water. *Read labels carefully.*

General Enzyme Guidelines: Eat as much fresh raw fruit, vegetables and protein as possible. Chew thoroughly so that saliva enzymes can be mixed with food for better utilization. Eat slowly. Avoid large quantities.

Dr. Leo V. Roy's program is designed to rebuild total health and thereby uproot and cast out infectious wastes causing arthritis distress. When you correct your entire metabolism, then the healing process can begin.

REDUCES SWELLING IN FOUR DAYS

Factory foreman Fred O. began to develop more than painful arthritis stiffness of his fingers. He experienced swelling. His knuckles became so "water-logged" that he could hardly move his hands. He found it painful to hold an ordinary ruler in his hand. He despaired, thinking that he was being made an invalid in the prime of his life. He discussed the problem with an examining nutritionally aware physician who outlined a basic enzyme-energizing program, similar to the one described above. Fred O. followed it with hope for relief. It came sooner than he expected. Within four days, the swelling was markedly reduced. He could use his hands with the flexibility of a youngster. Two days later, the arthritis-like affliction was gone. Fred O. follows the basic guidelines and is able to enjoy healthy hands and freedom from arthritis.

THE RAW FOOD "ENZYME-ENERGIZING" PROGRAM
FOR ARTHRITIS RELIEF

Once your body is given a supercharging of enzymes, these catalysts are able to set off a series of biological reactions within your endocrine glandular system so that an outpouring of hormones can transport important cortisone to your body sites where a natural protective action against arthritis takes place.

Raw Food Enzymes Strengthen Bones. These raw food enzymes distribute cortisone and other related hormones to your bone and fibrous connective tissues where strengthening is required. This creates resistance to inflammation and swelling. The deposition of enzyme-distributed cortisone will also offer a cooling effect, as well as matrix (intercellular substance or cartilage) strengthening. This action builds natural immunity to arthritis, stems the onset of arthritis and helps reverse its progress. Eaten regularly, raw food enzymes can strengthen your skeletal structure so you can free yourself from arthritic pain.

Arthritis Healings Reported. Searching for a lasting healing, a British physician used a raw food enzyme program that did more than relieve swelling, reduce inflammation and restore flexibility to formerly gnarled limbs. The raw food enzyme program promoted a lasting healing!

Dorothy C. Hare, M.D., reporting to the medical journal, *Proceedings of the Royal Society of Medicine*[3], explains that she supervised a number of her arthritic patients on this raw food enzyme diet at the Royal Free Hospital in England. There were many, many healings reported.

---

[3]*Proceedings of the Royal Society of Medicine,* "Raw Food Program For Arthritis," London, England, Vol. 30.

This program can also be followed at home. It is:

Breakfast. Apple porridge made of grated apple, soaked raw oatmeal, grated nuts, cream, fresh orange, tea with milk and cream.
Mid-morning. Tomato puree with lemon.
Dinner. Salad of lettuce, cabbage, tomato, root vegetables, oil-based salad dressing, mixed fruit salad, cream.
Teatime. Dried fruits, nuts, tea with milk and cream.
Supper. Fruit porridge, prune, apricot or apple salad with dressing.
Bedtime. Orange-lemon juice in boiled water.

Basic Program: For two weeks, the entire diet consisted of the preceding enzyme-rich foods. Then, certain cooked foods (vegetable soup, one egg daily, two ounces of meat, two ounces of bread, butter, cheese and milk) were added to the basic raw foods.

Guidelines. No salt was used for either raw or cooked foods. The dried fruits and raw oatmeal were soaked in water. All vegetables were shredded. Nuts were either whole or chopped. All food was prepared fresh for each meal. Servings were attractive.

Majority of Arthritis Patients Healed

Results: Eight out of the twelve patients showed much healing of arthritis *within four weeks.* (Two were helped but then went into a relapse; two showed no change at all.) Of the eight patients, seven showed continued improvement after going home.

Enjoys Freedom from Pain

Elizabeth P., at age 46, had a four-year history of pain and swelling. She had been bedridden for ten weeks. She was diagnosed as having fluid in both knee joints. There was painful swelling in her other joints. Elizabeth P. was put on this enzyme energizing program. She experienced so much relief that she could get out of bed and start to do regular housework in a short time. The pain, swelling and water-logging of tissues now started to subside. Thanks to the hormone stimulating effect of the enzymes, Elizabeth P. was no longer a chronic invalid. She had freedom from arthritis pain and felt healed.

Why the Enzyme-Energizing Raw Food Program Is Effective.

Dr. Dorothy C. Hare explains, "Nearly all of the arthritics were able to enjoy freedom from arthritic pain on this raw food diet. I attribute its success to the absorption of the unaltered solar energy (enzymes) of plant life... Science has so far revealed nothing... of this occult solar energy, as something apart from vitamin and chemical constituents of food." This suggests that nature-created enzymes in the raw foods are able to set off biological reactions that *correct the metabolic error* that has led to degenerative arthritis.

SOOTHE ARTHRITIS WITH IMPROVED
ENZYME-ACTIVATED DIGESTION

Activate your digestive powers with the use of enzymes and you will be able to help soothe and erase arthritis distress.

Simple Daily Program: Eat a freshly prepared *raw* salad of seasonal vegetables *before* each of your two or three main meals. Finish with a freshly prepared *raw* salad of seasonal fruits *after* your meals. *Benefit:* The raw vegetables start digestive and metabolic enzymes working so that ingested foods will be better dissolved and nutrients assimilated. The raw fruits tend to break down tougher ingested foods (meats and meat products) so that nutrients can be removed and used for the rebuilding of your cellular structure.

This simple daily program will help keep your enzymes active-alert throughout your lifetime. You will then be able to shield yourself against arthritis and other ailments related to poor metabolism.

Enzymes and Nutritive Assimilation. You may be eating wholesome foods but their nutrients are weakly or inadequately being metabolized because of weak enzyme power. This results in a deficiency, despite nutritive intake. Enzymes are needed to metabolize the nutrients.

Caution: If you are enzyme weak over a period of time, you lower your body's natural resistance to infectious ailments. Bacteria, viruses and parasites may now flourish and multiply. Defenses are lowered. Tissue and cartilage breakdown is the forerunner of arthritis.

Secondary arthritis is frequently traced to an enzyme deficiency because there is a weakness in the absorption, digestion and metabolism of essential nutritive factors.

Change Is Slow. This subclinical deficiency brings on slow and subtle changes in your bones and joints over a period of time. Gradually, you develop a degenerative bone and joint condition. Bit by bit, you find yourself weakening and growing more fragile.

That is why you should look to enzymes *at the start* to help guard against the "sneaky" slowdown in metabolism and slow onset of arthritis.

The Pain-Relieving Enzyme Tonic

Problem: Bernice G. was troubled with more than just joint stiffness. She had excruciating pain in her knuckles and hip joints that began to spread down to her knees. This arthritis-like pain made her wince whenever she had to do bending or twisting in even ordinary gestures. She wanted to have pain relief at any price!

Suggested Healer: Her pharmacist noted that chemotherapy and prescribed medicines only made her worse, especially if she eased on taking them. She developed side effects such as dizziness and skin rashes from such patent remedies and sought a natural method of pain relief. The pharmacist suggested she try a simple *Pain-Relieving Enzyme Tonic.* She did. Bernice G. drank this simple all-natural tonic twice a day. Results? Within four days, all pain was gone. Now she became as flexible as an athlete. She joked she could now be an acrobat!

Pain-Relieving Enzyme Tonic. To a glass of freshly prepared pineapple juice, add one teaspoon of brewer's yeast, one teaspoon of desiccated liver, one teaspoon of alfalfa seeds. Blenderize for just one minute. Now sip slowly. *Suggestions:* Drink one glass in the morning, another glass at noon and a final glass in the evening. Do this every single day until pain relief is enjoyed; usually, this is experienced after two days.

Enzyme Benefits: When you drink this tasty and effective *Pain-Relieving Enzyme Tonic,* a series of "instant" reactions occur to give you quick and welcome relief. The *bromelain* enzyme in the pineapple juice takes up Vitamin C and combines it to pantothenic acid, the B-complex vitamin found in the brewer's yeast, desiccated liver and alfalfa seeds.

Almost immediately, the enzyme will create the Krebs Cycle, a process by which acid by-products of energy manufacture are transformed into water and carbon dioxide for removal from your body. This causes a speedy waste removal so that pain-causing lactic acid can be washed out of your muscle and joint tissues.

Tonic Guards Against Arthritis. The high concentration of pantothenic acid in the tonic is needed for new bone formation and to protect against excessive calcification in your joints. A deficiency of enzyme-activated pantothenic acid may cause your cartilage to become calcified and hardened and this predisposes you to arthritis. Therefore, the simple tonic can help protect against this arthritis-causing deficiency.

Special Benefit. The *Pain-Relieving Enzyme Tonic* offers a high concentration of pantothenic acid which stimulates the production of adrenal cortical hormones such as *cortisone.* This is *natural cortisone* which is needed to help ease or eliminate arthritic pain. It is your body's *own* cortisone that is activated by enzymes in the *raw* pineapple juice to provide this soothing balm and relief from swelling, calcification, accumulated deposits and inflammation.

Say Goodbye to Arthritis with Enzymes

You do *not* have to learn to live *with* your arthritis. Instead, you can learn to live *without* those agonizing pains with the help of enzyme activation. This correction of the biological error will help you say goodbye to this unwanted and undesirable result of weak metabolism. Follow a raw food enzyme program and restore youthful flexibility to your joints…and entire body and mind, too.

MOST IMPORTANT:

1. Follow the 7-step Dr. Robert Bingham program for clearing up arthritis.

2. Susan K. experienced relief of swollen fingers and arthritic pain within six weeks after following the doctor's program.

3. For a specific arthritic condition, note Dr. Bingham's suggestions for enzymes together with individual vitamins.

4. A Canadian doctor has prepared a lifetime "Freedom From Arthritis" Enzyme Program easily followed in your own home.

5. Fred O. reduced arthritis stiffness and swelling on this Canadian doctor's program within four days!

6. A British doctor's raw food "enzyme-energizing" program provided welcome arthritis relief for a set of formerly afflicted patients.

7. Elizabeth P. followed the British program for almost immediate relief after suffering for four years with painful swelling. Formerly bedridden, she got up and started to do housework after being on this enzyme program for a short time.

8. Bernice G. prepared an easy *Pain-Relieving Enzyme Tonic* that actually boosted *natural cortisone* and offered almost instantaneous freedom from swelling, calcification, accumulated deposits and inflammation.

CHAPTER 4

The Skin-Rejuvenating and Hair-Growing Power of Enzyme Catalyst Foods

Skin regeneration and hair health depend upon the vitality of microscopic substances that lie beneath the surface of your body covering.

These are powerful enzymes that perform two basic functions: (1) clean away accumulated toxic wastes and prepare them for elimination; (2) stimulate the health of the follicles in your scalp so that they are energized to improve natural and vigorous hair growth.

With the use of enzymes, you can improve your appearance, help firm up wrinkles, wash away acne, clear up blemishes and improve the thickness and quality of your hair. You will then be able to look and feel much younger than your biological age.

Let us see how you can use enzymes in everyday foods to help alert your sluggish responses to stimulate internal reactions to improve your skin and hair.

I. HOW ENZYMES REJUVENATE YOUR SKIN

Skin Needs Enzymes. Your *epidermis* (outer skin) needs to be energized with enzymes in order to continue its ceaseless process of shedding dry dead cells and toxic wastes. Without enzymes, or if your *epidermis* has a weak enzyme supply, this function becomes sluggish. Accumulated wastes cause a dehydration and erosion effect. This paves the way for blemishes, wrinkles and premature aging signs.

Your *dermis* (inner skin) lies directly beneath your epidermis which contains millions of capillary blood and lymph vessels, nerve endings, sebaceous and sweat glands that depend almost entirely upon enzymes so that they can nourish your skin and deliver waste products for removal to your sweat glands. If your skin is enzyme-weak, then your *dermis* becomes weak. New cells cannot be formed. The small moisture reservoirs lying right in your *dermis* become "thirsty" or parched. This causes dehydration which then is the forerunner for wrinkling as well as skin aging. *Enzymes will energize your dermis to create this cleansing and nourishing rhythm that is the key to a forever young skin.*

Enzymes Create Fibroblasts: Key to Youthful Skin

In the superficial layers of your dermis are found special cells called *fibroblasts.* These little-known "youth workers" have the biological task of continuously making and regenerating connective fibers throughout your entire body. *Fibroblasts* derive this energy and the raw materials for cell-tissue regeneration from enzymes. If you have an ample amount of enzymes available (externally and internally) then your *fibroblasts* can create this perpetual rejuvenation. You will then glow with youthful firmness. You will have a skin that is healthfully clean and free of blemishes and aging signs.

Putting it simply, feed enzymes to your *fibroblasts* and they will then create the cell-tissue regeneration that will give you "forever young" looking skin.

Enzyme Salad for Schoolgirl Complexion

Troubled with wrinkles, crow's feet, unsightly circles under her eyes, Stella N. asked a dermatologist for help. She was advised to eat one or two raw vegetable salads daily. She was also advised to eliminate sugar, salt, harsh seasonings, fried and fatty foods. (They slow up enzyme process.) Stella N. prepared a simple enzyme salad and enjoyed it before each of her main meals, daily. Within five days, her skin did more than just clear up. The wrinkles and folds and circles went away. She now had an unusual glow. She had an enviable schoolgirl complexion. She was "young again."

Enzyme Salad. On a bed of shredded cabbage, have either chopped or shredded carrots, tomato slices, radishes, sweet red pepper slices, celery chunks, cucumber slices. Make a dressing of simple apple cider vinegar and your favorite oil. Sprinkle a bit of onion and/or garlic powder (not salt), according to taste.

Suggestion: Chew very thoroughly. This activates your salivary enzymes and alerts your digestive juices to metabolize the ingested salad more effectively. There will be speedier absorption if you start with better chewing of your Enzyme Salad.

Benefits: When digested, your *fibroblasts* will take the enzymes and nutrients from the salad, then form collagen or elastin fibers and redistribute them in the surrounding dermal connective tissue. Your fibroblasts use these enzymes to help guard against skin tone loss, or sagging of elasticity which is the forerunner of aging skin. Available in the high-powered foods of the preceding *Enzyme Salad,* these substances are able to perform two unique skin-rejuvenating actions:

1. Enzymes stimulate formation of collagen to induce moisturizing (hydration) of your dermis.
2. Enzymes also work with fibroblasts to manufacture elastin which then creates more youthful elasticity of your skin.

Suggestion: Plan to eat the *Enzyme Salad* at least twice daily to provide powerhouses of enzymes for your *fibroblasts* to create this skin rejuvenation action.

How to Protect Enzymes Against Wind Abuse

You can protect your enzymes and "save face" when winds tend to abuse your skin, making it raw, rough and weatherbeaten. Constant summer or wintry winds can cause overactivity of the oil glands beneath your skin surface and this leads to enzyme evaporation. To understand, note this guideline:[1]

Reason: Your skin's surface is kept soft by water from within, released slowly through an underlayer of the skin. Your natural oils control the release of this necessary youth-maintaining, enzyme-created moisture. However, too much wind removes these natural protective oils and your skin loses its youthful, moist glow.

Simple Home Remedy: Each morning, apply a light but thorough coat of petroleum jelly to your skin to form a fine, translucent shield against drying winds.

---

[1]*Parker Natural Health Bulletin,* West Nyack, New York 10994. Vol. 9, No. 13, June 18, 1979. Available by subscription.

Suggestion: If you're going to be outdoors all day, keep re-applying the petroleum jelly at intervals. Hands, cheeks and chins will benefit from a light, comfortable coat of this softener.

Home Enzyme Activating Remedies for "Total Youth" of Your Skin

In the privacy of your own home, using enzyme-containing foods and other natural ingredients that stimulate enzyme activation to provide skin-rebuilding nourishment for your *fibroblasts,* you can help rid yourself of winkles and blemishes, often overnight. Here is a list of such easy-to-follow skin-regenerating remedies.

1. Egg white is a stimulating basic facial and skin toner. Just clean your face well and rub with egg white. It will tighten on your skin and require plenty of rinsing, but it leaves your complexion rosy, glowing, immaculate.

2. If you do use powder or make-up base (the ladies, that is!) make sure to remove it well before bedtime. Stale make-up clogs your pores and actually chokes your enzymes to death. Baby oil is a good cleanser. It also tends to moisturize your skin cells and satisfy the thirst of your *fibroblasts.*

3. An enzyme paste mask is excellent for tightening up enlarged pores. Make a paste of oatmeal or cornmeal by mixing with a little water. Spread on your face. Let remain for at least 15 minutes. Rinse with warm and then cool water. Helps scrub the pores free of debris and then enzyme-nourishes your *fibroblasts* to make more collagen that will tighten and close the pores.

4. For an oily skin, blend avocado to a paste and apply generously. Let remain 15 minutes. Rinse with warm and cool water. Enzymes will use nutrients and amino acids from the avocado to help stimulate renewed cell growth and also help balance oil levels in your reservoirs so that there is an even distribution, rather than an overflow.

5. Your eyelids (they're skin, too!) can show wrinkling from enzyme starvation. To refresh, soak ordinary gauze in freshly squeezed orange juice and apply to your lids. The orange enzymes will help stimulate new collagen growth and firm up the wrinkled folds. Keep the gauze wet with juice and let it remain on your lids about thirty minutes.

6. Cucumber enzymes are invigorated by that vegetable's high potassium and phosphorus content. In combination, these nutrients help stimulate sluggish *fibroblasts* to promote cleansing

and firming up of your skin. *Suggestion:* Rub raw cucumber slices all over your face. Or, rinse your face with cool water in which you've mashed some cut-up cucumber. The enzymes and minerals will create a powerhouse of vitality so that your *fibroblasts* wake up and create new skin cells in a matter of moments.

7. Try any of these enzyme activating remedies for dry skin: rub your face lightly with ordinary olive oil. Try the juice of honey-dew melon as a mask. Or mix the mashed melon with petroleum jelly and spread on your face. The raw enzymes will become moisturized and energized and your skin cleansing-renewing process will be all the more activated very shortly.

How to Catalyze Your Metabolism for "New Skin" in Moments

To protect against aging, you need to plump and firm up your body's supplies of elastin. With sufficient enzyme action, you can alert or catalyze your metabolism so that your body fibers become more invigorated in tone and elasticity. You can then protect against problems of wrinkles and sagging skin and age-appearing blemishes.

Here is a set of home programs to follow. Each is aimed at either introducing enzymes to your skin externally, or stimulating your enzyme production under your skin. They reportedly[2] can help "put roses in your cheeks."

BERRY YOGURT MASK

Combine two tablespoons of plain yogurt with one tablespoon of strained strawberry juice (made from fresh or frozen strawberries.) Stir by hand in a small bowl until the ingredients are blended. Apply the mixture to your face with your fingertips, and let it soak in for up to 30 minutes. Then splash off with tepid water followed by cool water. Your face will look and feel clean and youthfully alert!

HONEY AVOCADO MASK

Combine two tablespoons of honey with one tablespoon of mashed avocado. Add one whole egg (or two egg whites) and mix the ingredients by hand or in a blender. Apply the mixture to your face with your fingertips. Let it remain for up to 15 minutes. Then rinse with warm water while you gently massage your skin, using a

[2]*Parker Natural Health Bulletin,* West Nyack, New York 10994. Vol. 8, No. 10. May 8, 1978. Available by subscription.

circular motion. This helps correct oily skin and tightens up your pores.

OLD-FASHIONED CLAY MASK

You can concoct an old-fashioned clay mask by combining two tablespoons of kaolin (fine white clay available at most pharmacies) with 1/4 teaspoon of peppermint extract. Add one tablespoon of water. Thoroughly blend the ingredients. Apply the mask to your face. Let it remain for 15 minutes. The clay does not dry on your face, so continue to massage your skin for several minutes. Rinse with tepid water followed by cool water. This treatment should help remove debris, tone up your skin surface and make your face feel clean and youthfully glowing.

Young Skin in Two Days

Edna C. was troubled with oily skin. Unsightly oil clogged her pores, gave rise to blackheads. She also developed furrows and wrinkles that made her look haggard all the time. Her cosmetician told her about a simple mask that would do more than remove excess oil. The mask would penetrate the pores, send forth powerful minerals that would be transported by enzymes to the *fibroblasts* that would rebuild the elastin and collagen so that "new skin" would be the reward. Edna C. tried the mask. Within two days, her oil problem was over. Now she was rewarded with clean and wrinkle-free skin. She looked youthful (instead of haggard) all the time!

OIL-REMOVING MASK

Before You Begin: Keep yourself very relaxed so that your facial muscles will not become tight and strained.

How To Make Oil-Removing Mask: Combine 1/2 cup of fuller's earth (available at pharmacies and health food stores) with one tablespoon of alcohol and two tablespoons of witch hazel. Spread this mask over your face, rubbing it *very gently* into those areas that are particularly oily. Let the mask remain until it is completely dry. When it cracks, remove it with cold water and gentle rubbing. You'll discover much of the excess oil is gone, the pores closed. Soon, blemishes will be gone. Your skin emerges with smooth firmness. *Reason?* Enzyme-dispatched minerals from the fuller's earth have nourished your *fibroblasts* so they can knit and bind the fragile collagen and give you "new skin" from within.

Plan to use any of these masks as often as possible. They work speedily and effectively. They help create cleansing actions to dispose of toxic wastes and thereby build and rebuild healthy skin...at any age...in a short time.

15-Minute Skin Cleanser

Just hold your face over a bowl of steaming water for 10 to 15 minutes. Then splash with tepid water and towel dry.

Benefit: The steam opens up pores and helps steam out dirt particles and cell-abrasive wastes. Your skin becomes softer. The steam also tends to moisturize your enzymes which then deposit much needed liquids upon your thirsty cells and tissues. This helps plump and firm up your cells and guards against wrinkling.

Enzyme-Catalyst Tonics to Wash Away Acne

Cause Of Acne: An excessive release of sebum, an oily substance discharged by the *sebaceous* glands (directly beneath skin surface) forms clogged pores. This condition leads to acne. While it may appear anywhere on the body, acne proliferates on the face where more sweat glands are located. It is not an affliction exclusively for the young. In middle-aged folks, *seborrhea,* as it is known, affects areas of the face and body. It is usually called *rosacea* but it is the same problem; namely, acne. A malfunctioning of the sebaceous glands leads to the affliction.

Enzyme-Catalyst Requirement. To stabilize internal secretions and, even more important, to cleanse away debris that causes toxic waste accumulation, enzyme catalysts are required.

Basic Benefits: When you drink enzyme-catalyst tonics, the assimilation of these ingredients is accelerated and they are absorbed speedily into the bloodstream. The enzymes stimulate the *fibroblasts* to form more elastin and collagen connective fibers and at the same time to create a "scrubbing" action so that toxic wastes can be broken up and then eliminated. This creates internal revitalization and a "washing away" of acne and other skin blemishes.

Fresh, raw enzyme-catalyst tonics can create this skin clearing process. You can easily make them right in your own home. They are easily prepared. They work swiftly. Try these acne-washing programs.

Papaya-Pineapple-Orange Juice. The papain-bromelain enzymes are exhilarated by the Vitamin C of the orange juice and work harmoniously to cleanse away dead matter and balance normal sebum release.

Carrot-Cabbage-Green Pepper Juice. Enzymes use the Vitamin A (in the form of carotene which is metabolized in usable form) together with the minerals to help strengthen the connective tissue and bring about a sloughing away of accumulated toxic wastes.

Apple-Pear-Grapefruit Juice. Enzymes combine with minerals and Vitamin C to create almost "instant" scrubbing of cells, refreshing them and helping to cleanse away pores so that acne is reduced and eliminated speedily.

Lettuce-Turnip-Celery Juice. High concentrations of cell-washing vitamins and minerals are energized by enzymes and create a helpful cleansing reaction.

Tomato-Cucumber-Radish Juice. Enzymes join with the high mineral content to create a more even distribution of sebum and also help bring about improved cleansing of the cells and tissues.

Turnip Leaves-Dandelion Leaves Juice. The high potassium content in these leaves creates a strong alkalizing reaction. This helps reduce excessive acidity, which is often the cause of waste accumulation traced to clogging of tissues and cells. The high magnesium content of the dandelion leaves is energized by enzymes to blend with the calcium of the turnip leaves and strengthen your cellular structure. You'll develop clearer and more youthful skin in a short while.

Carrot-Lettuce-Watercress Juice. Enzymes will seize the rich store of the sulphur mineral and join it with phosphorus and potassium and use these minerals to create a catalyzing action on the accumulated wastes clogging the pores. These wastes are speedily dislodged and washed out of the system. Internal cleansing will lead to external healing of acne and other blemishes.

Father and Son Grow "New Skin" in Three Days

In his middle years, Albert D., Sr., developed acne. In addition, his skin became so dried that wrinkles and creases developed like indented grooves. At the same time, his young son, Al D., Jr., became the victim of stubborn acne pits as well as blackheads and unsightly blemishes. Both of them sought help from a university biologist who specialized in dermatological problems. He told them of programs whereby intake of a variety of fresh raw juices could help wash away debris; but more important, create and regulate hormonal secretions. He suggested they try a very simple program:

Enzyme-Catalyst Juices. For five days, take no other liquids (except water) but a variety of desired fresh raw vegetable juices. Father and son were to consume as many quantities as they could comfortably accommodate. They followed this simple program. Results? Within three days, their acne and blotches almost vanished. By the end of the fifth day, the father's skin was now normally moisturized and "plumped up" so that wrinkles had gone away. He quipped that he looked as young as his son because both had "new skin."

Benefits: Ordinarily, the connective fibers in the skin may deteriorate if improperly nourished with essential vitamins and minerals. Enzymes are needed to use these nutrients to bring about a continual regeneration of these fibers. Enzymes create a catalyst action by which they dissolve and destroy aged fibers and then stimulate the *fibroblasts* to create *new* ones. The enzymes in the raw juices worked swiftly to create this healing reaction.

Caution: If there is a sluggish metabolism and a reduction in essential fibrillar rejuvenation, then decay remains in the *epidermis* (outer skin) and *dermis* (inner skin) where it causes congestion of the dermal connective tissue. This leads to acne as well as furrows and winkling. Aging skin occurs.

Enzyme-Catalyst Solution. Have an adequate, if not superadequate quantity of enzymes available from fresh raw fruits and vegetables and their juices. These enzymes offer a reservoir of raw materials (nutrients) which nourish the entire dermal tissue. These enzymes also use a catalyst action to wash away the "old" and replace with the "new" cells and thereby help you glow with youthful looking skin.

II. HOW ENZYMES HELP GROW HAIR

Scalp Needs Hair-Growing Enzymes. Hair grows out from follicles (or hair roots) located at the bottom of the layer of your scalp skin. Along the tubes from which the growing hair shaft emerges are situated several tiny oil glands. If these oil glands are sluggish, hair growth and replacement is weak; hair, itself, is brittle. Dandruff may result. If these oil glands are overactive, then hair (and skin) become too oily or greasy, attract dirt, invite clogging and scalp problems. Therefore, the goal is to have *healthy* oil manufacture; with the use of a natural "balancer," this can be accomplished. Such a substance is the enzyme. By making enzymes available to the scalp, the nourishment of this skin covering is helpful in growing healthier hair. Loss can also be controlled, if not averted.

Food Enzymes Create Hormone Hair-Growing Process

An ample supply of food enzymes will help energize sluggish hormone responses. In particular enzymes perform a unique hair-growing reaction.

Balance Hormone Output. In most cases of hair loss, there is an excess of *androgen,* a male hormone, stored in the scalp. This excess saturates not only the hair shaft but the skin of the scalp, too. A problem is that stored up *androgen* eventually breaks down chemically. It releases a by-product known as DHT. This mischief-making waste product tends to clog the hair follicle, causing it to degenerate until it cannot grow hair. This is one basic cause of hair loss. Therefore, the answer to the problem is to find a substance that will do more than balance androgen hormonal output; one that will break down and wash away the androgen so it cannot destroy hair follicles.

How Enzymes Catalyze Androgen Excess. When your bloodstream transports adequate enzymes to all body parts, including the skin of your scalp, a unique reaction occurs. The enzymes tend to dissolve debris clinging to and filling hair follicles. This helps free the follicle and give it a chance to grow hair. The enzymes catalyze or break up androgen, and then loosen the subcutaneous tissue surrounding the hair follicle. Almost at once, your scalp's capillary system enjoys more blood circulation. Now nutrients are able to nourish the follicles and hair has a chance to grow.

Enzymes, Hormones, Scalp Cholesterol. Enzymes (both external and internal) tend to break up and dissolve accumulated androgens and DHT. But they perform a unique action. They get to the "root" of the hair loss problem — the substance that causes an excess of androgens and DHT in the first place — cholesterol. Located within your scalp tissues, cholesterol appears to bring about production of both androgens and DHT, the two substances that clog and destroy hair follicle tissues. Enzymes have this "search and cleanse" action. They strike at cholesterol, break it down, and help wash it out of the scalp. Now the "cleansed" hair follicles have an opportunity to grow new hair.

New Hair on Bald Spots. by eliminating the cause of balding (androgens, DHT and cholesterol), enzymes help nourish the hair roots that have deteriorated. Now, many hair follicles have stopped creating hair because of the choking effect of DHT. They need to be regenerated. The follicles consist of cells and tissues that are energized and nourished by enzymes. Once the DHT is removed, then cell growth and reproduction of new cells can take place. By eliminating or neutralizing the anti-hair substances, the scalp has a chance to grow new hair on formerly bald spots.

FRUIT JUICES + GRAINS = HOPE FOR HAIR REGROWTH

Fresh fruit juices combined with whole grains offer a unique combination that promotes scalp health and offers hope for new hair regrowth.

Grapefruit Juice + Wheat Germ + Bran = Nutrient Powerhouse. Blenderize these three foods. Drink several freshly prepared glasses daily.

Benefits: Enzymes in the grapefruit juice take up its Vitamin C and use it to create collagen, the connective substance located in your hair follicles. More important, the enzymes will also take the B-complex vitamins (biotin and niacin) and also cystine (an essential amino acid) and use this combo to cleanse the scalp and stimulate new hair growth.

Hair-Growing Power of This Combo. The nutrients in this combo create these benefits, almost immediately:

Vitamin C. Manufactures collagen to provide needed cellular regeneration of the scalp hair follicles.

Biotin. Creates a favorable environment for hair restoration; also is needed to help wash out DHT (dihydrotestosterone) and cholesterol deposits from hair follicles.

Niacin. Acts as a scalp blood vessel dilator; boosts the flow of nourishing blood to your scalp so hair follicles can feed on nutrients that prompt hair growth. This circulation-boosting process aids polymer (molecular) penetration of your hair's porous structure. Now your new hair is able to be strengthened as it is able to regrow.

Cystine. This essential amino acid is propelled by enzymes to your scalp. It has a unique value. It is the most essential building block of *keratin,* a protein substance which makes up much of your hair. Enzymes use cystine with the other nutrients of this juice combo to nourish the cell nuclei of your hair shaft. It also acts as a protective coating of your hair.

A Combination Is Most Effective

These nutrients appear to work more effectively in this special combination. The grapefruit juice offers enzymes plus Vitamin C which then dispatch the biotin, niacin, and cystine to your scalp and hair molecules. They quickly begin (1) cleansing of DHT and follicle-destroying wastes, (2) nourishment of the hair follicle and shaft and (3) strengthening of the new hair. It is the *combination* that is most effective.

Hair Loss Halted, New Growth Starts to Appear

Stephen Y. was troubled with regular hair fallout. Daily, he had more hair on his brush than ever before. Scalp massage offered some help but he was still bothered with thinning hair. He obtained help from a hair-growing specialist. He was told to try the aforedescribed enzyme hair-growing tonic. Within three days, Stephen Y. saw hair loss diminish. At the end of five days, not only did this enzyme catalyst tonic halt hair loss, but it started new growth. Within seven days, he had more hair to brush than he remembered...and hardly any came off! He drinks this tasty tonic daily and enjoys his "rediscovered" head of hair.

How to Be Good (and Healthy) to Your Hair

To nourish your "crowning glory" you need to keep your hair in good enzyme active condition. Here are several easy-to-follow programs that put new life (and hair) into your scalp's tresses:

1. A weekly shampoo and daily brushing are important. Whether you have problem hair or naturally beautiful hair, brushing daily will make it more shiny and healthy right down to your roots. A hundred strokes a day will do it...well worth the time. Good exercise, too.

2. *For Blondes Only:* Rinse your hair in a brew made of one half cup of camomile flowers and two cups of comfortably hot water. Or use grapefruit juice to rinse.

3. *For Brunettes Only:* Try a brew of rosemary steeped in water and drained. Makes for great highlights and sheen.

4. *Oily hair?* Beat two egg whites until stiff and then apply this pure protein to your scalp with an old toothbrush. Let the whites dry and then brush thoroughly. Shampoo off.

5. *Dry hair?* Rub castor oil into your scalp at bedtime and shampoo in the morning (reserve an old pillowcase for oil-treatment use). Treat twice a week for a few weeks, then once every two weeks. It will make your hair shiny and more healthy. *Note:* If you don't want to leave the oil on all night, rub the castor oil well into your scalp and steam it in by pressing a hot towel on your head. Shampoo well.

6. *Dandruff?* Try mixing equal amounts of vinegar and water, part your hair and apply well to your scalp with cotton before shampooing.

7. *Hair nourishment?* Beat one raw egg lightly with a fork and rub well into your scalp, using it as a substitute for shampoo. Prime source of enzymes, and also important cystine. Rinse well with warm water.

8. *Natural Setting Lotion.* You can enzyme-protein enrich your hair by combining ordinary milk with a little bit of lemon juice. Use it as a setting lotion. It adds body and luster to your hair, nourishes your scalp and leaves no flakiness or odor. *How To Use:* After shampoo and rinse, spray on the milk-lemon juice combo with an atomizer, comb through your hair and set as usual.

BASIC HOME CARE FOR YOUR HAIR[3]

You'll boost your enzyme catalyst cleansing and hair growing reactions by following these home care of the hair tips.

IF YOUR HAIR IS OILY

1. Plan to shampoo with a mild product at least every other day.

2. Wet your hair with warm water before shampooing. Work up a good lather, then rinse with cool water. This helps control the flow of excess oil.

---

[3]*Parker Natural Health Bulletin,* West Nyack, New York 10994. Vol. 8, No. 16. July 31, 1978. Available by subscription.

3. A lemon or vinegar rinse will help cut through and remove excess oil. Add a small quantity of enzyme-rich vinegar or lemon juice to very cool water and splash it over your hair.

4. Very vigorous brushing can increase the flow of excess oil, so brush gently with a natural bristle brush. Aim for about 50 brushstrokes in the morning and 50 brushstrokes at night.

5. Before brushing, pack strips of cotton between the rows of bristles to help absorb oil.

IF YOUR HAIR IS DRY

1. Brush frequently to stimulate oil flow and to distribute the oil from your scalp to the ends of your hair. Aim for about 75 brushstrokes in the morning and 75 brushstrokes at night.

2. Use a mild shampoo once or twice a week. You might use castile shampoo, egg shampoo or any natural soap that is labeled "for dry hair."

3. Lather and rinse only once. Your hair is already too dry and a second shampoo would only wash out more oil.

4. Massage your scalp to stimulate sluggish oil glands. Place your fingertips firmly against your scalp. Rotate them gently so that you feel your scalp moving slightly. Be careful not to rub too hard, because the friction may cause hair loss. Keep moving your fingertips until you've covered your entire head. Repeat the massage twice a day for about five minutes.

5. Treat your dry hair to the time-tested hot oil remedy. It's simple but very effective. Warm about half a cup of olive oil and use cotton pads to dab it all over your scalp. Now wrap your head in one or two hot hand or dish towels. (You can warm towels by dipping in hot water and wringing them out or by wrapping them dry around your head and warming them under a hair dryer turned on low.) Keep your head wrapped in the hot towels for about 15 minutes. Then remove towels and shampoo as usual. This hot oil remedy, repeated three times a week, helps moisturize your dry hair and scalp.

6. Protect your hair against the drying, bleaching effect of the sun and wind. Always wear a scarf or hat when outdoors in the sun. If you've been swimming in salt water or a chlorinated swimming pool, be sure to rinse your hair well under a shower. Salt has a corrosive action on the hair; chlorine is a bleach that can cause drab or brittle hair.

With improved skin and hair enzyme nourishment, basic home care, you can help rejuvenate your body envelope and boost growth of your scalp tresses.

SUMMARY:

1. Enzymes create fibroblasts which give you youthful skin.
2. Stella N. used an enzyme salad to develop a schoolgirl complexion within five days.
3. Use any (or all) of the enzyme catalyst programs for "new skin".
4. Catalyze your skin and help it glow in moments with any of the home programs outlined.
5. Edna C. used an oil-removing mask that cleansed her pores, washed away blackheads and smoothed out wrinkles, all within just two days.
6. Drink the enzyme-catalyst tonics to wash away acne.
7. Albert D.,Sr., and Al D., Jr., a father and son duo, developed "new skin" within three days. They looked like brothers!
8. Enzymes and nutrients help grow hair when used in a simple fruit juice plus grains combo.
9. Stephen Y. halted hair loss and then caused new growth to appear within seven days on the enzyme program, especially with the special tonic.
10. Improve general hair and scalp health with the set of easy-to-follow home programs.

CHAPTER 5

How Enzymes Wake Up Your Glands for

Happy Hormone Health

The health of your skin, the flexibility of your joints and muscles, the energy required to meet responsibilities of daily living, the metabolism of ingested foods, the very foundation of youth — all depend upon one group of important organs, *your glands.*

These internal "fountains of youth" are able to release health building substances called *hormones.* These secretions regulate and also rule just about every activity of your body and your mind. An adequate amount of hormones will help keep you youthful and healthy. An excess can cause erratic physiological and emotional responses. A deficiency can cause aging and the onset of gradual deterioration of the body. The goal is to keep your glands healthy so they release a *balanced amount* of hormones every moment of your life.

ENZYMES: BIOLOGICAL MESSENGERS TO YOUR GLANDS

An abundant availability of enzymes in your system will stimulate or catalyze the actions of your glands to release an adequate amount of needed hormones. Without enzymes, your glands become sluggish and are unable to become sufficiently invigorated to distribute these essential hormones. Now the entire body starts to weaken and illness may strike. Prolonged enzyme deficiency can cause serious consequences.

Raw Food Boosts Healthy Hormones. A higher intake of fresh raw food will help stimulate your sluggish glands with enzymes and prompt them to release an adequate quantity of healthy hormones. Enzymes in the raw food are the energizers for your glands. Without them, your glands (and your body) would be in a state of decline. So enzymes control the power of your internal "fountains of youth" or your glands.

Selective Ability of Enzymes. Each enzyme acts *specifically* upon individual body glands. That is, one enzyme stimulates the release of insulin, another is involved in hormones controlling blood pressure, another stimulates your glands to improve your bloodstream. You need a *balance* of enzymes in order for *all* of your glands to be activated and alerted so they can release these "happy hormones." They are so called because your emotional health is strongly influenced by your enzyme-alerted glands. A deficiency of one enzyme means that the specific gland it acts upon becomes sluggish or sleepy. This can cause internal upset and vulnerability to ill health, so it is vital to be enzyme-nourished for complete glandualr alertness.

Biological Digestants. Enzymes act as small biological digestants which have the individual power to break down and dismantle protein, carbohydrate and fat, remove their essential components and use them for stimulation and regeneration of your body glands. Enzymes then "wake up" your sluggish glands with this nourishment; they do so by regulating the rate of biological reactions in terms of breaking down food and providing energy for your glands and also in building up needed raw tissue and cells of your glands. As biological digestants, enzymes are able to nourish your glands, keep them active and stimulate them to create healthy and happy hormones. Reward: *you* are now able to enjoy a healthier and happier lifestyle.

HOW TO ENZYME-ENERGIZE YOUR GLANDS

Let's take your body's individual glands and see how you can use enzymes to energize them and thereby discover a new joy of daily living.

Pituitary Gland

Location: This pea-sized gland is suspended from a short stalk at the base of your brain. It is called the "master gland" because its three lobes release at least nine known hormones.

Hormone Function: Various hormones stimulate flexibility of your muscles, regulate water balance, help control kidney function, even determine healthy skin color. Body size is also determined by healthy hormone balance. Pituitary hormones regulate output of your adrenal glands (discussed below) by releasing adrenocorticotrophic hormones or ACTH. This hormone is essential for healing a variety of ailments. To enjoy better health in daily living, the availability of ACTH as a natural internal healer is required. This can be possible through pituitary stimulation with the use of enzymes.

Enzyme-Energy Program. To start off your day, prepare a fresh raw fruit juice. Orange juice combined with grapefruit and pineapple juice and also cherry juice is supercharged with enormous amounts of Vitamins A and C as well as important minerals that are able to provide almost "instant energy" to your pituitary. Almost immediately, you will feel youthful muscular flexibility, improved skin color and more vitality. Enzymes in the raw fruit juice will use these vitamins and minerals to activate your pituitary to release the important energy-building hormones.

Health Decline Traced to Enzyme-Starved Gland

Saleswoman Helen S. felt increasing weakness as her workday wore on. She found it difficult to remain on her feet. She began to make frequent bathroom trips. Weight increased, even though her food intake was modest. Helen S. grew irritable upon the slightest provocation. She sought help from the store's registered dietician who suggested she stimulate her sluggish glands, especially her pituitary. She recommended a simple program. It consisted of a raw food plan for one day a week. Helen S. tried it. Overnight, she recovered. Next morning, she was filled with vitality, stopped going to the bathroom, and her waistline slimmed down. She was cheerful again, thanks to her happy hormone program.

Happy Hormone Program. Start your week off with just one day devoted to intake of raw foods and fresh raw juices — *no cooked foods.* You may have whole grain cereals (granola or no-cook natural variety) in milk for breakfast, but emphasize as many raw foods (fruits, vegetables, grains, seeds, nuts) as possible for this one day. Beverages should also consist of raw juices or, if desired, herb tea flavored with a bit of honey and lemon juice.

Benefit: Your digestive system propels enzymes to your pituitary to stimulate it to release hormones that metabolize carbohydrates (sugars and starches) and also improve calcium and protein assimilation. At the same time, the vitamins and minerals in the raw foods are used by enzymes to stimulate your pituitary to act upon your adrenals (part of a chain reaction) to release valuable ACTH. This is a disease-fighting hormone. It can help, as it did for Helen S. above, in preventing weight gain, water-logged tissues, nervous disorders. To ACTH-fortify your body is to strengthen your entire system and enjoy happy health...all over! You can do this with raw food enzymes.

Thyroid Gland

Location: It is a butterfly-shaped, two-part gland that rests against the front of your windpipe. It releases thyroxin, a hormone that stimulates the metabolism (or activity) of your many body cells and tissues.

Hormone Function: Thyroxin determines the rate at which your body burns up its foodstuffs and consumes oxygen. Too much thyroxin (hyperthyroidism) accelerates metabolism; too little (hypothyroidism) slows it down. This can cause excessive weight loss or excessive weight gain. In some situations, an erratic thyroxin release may cause rapid beating of the heart and pulse, fast breathing, development of goiter, eye bulging and extreme nervousness. The rate at which thyroxin is released is determined by the availability of enzymes in your system.

Enzyme Control Program. Enzymes need an essential trace mineral (so-called because its requirements are minute but its deficiency can cause serious upheavals) called *iodine.* Enzymes will use this mineral to control the cellular oxidation rate of thyroxin. Even the smallest shortage of iodine can cause disturbances. But even more important is that the *weakness* of enzymes can render adequate amounts of iodine insufficient. Therefore, you may have adequate iodine but weak or inadequate enzymes cannot distribute that iodine to your thyroid gland. A general guideline could be: enzymes *first* and iodine *second* for total thyroid health.

Thyroid Nourishing Tonic. You can wake up your sluggish thyroid with this delicious and easily prepared tonic. To one glass of tomato-celery juice add one-quarter teaspoon of kelp powder (made from seaweed, available in all health stores) and stir vigorously. Drink one glass of this *Thyroid Nourishing Tonic* just once a week. That's all. Immediately, your metabolic processes will seize the essential iodine and use it to invigorate your thyroid. You'll start to look and feel better in minutes.

From "Aged" to "Athlete" in One Day

Oscar Q. was troubled with bouts of excessive weight gain and then unnatural weight loss. He looked aged with his wrinkled skin and constant fatigue. He was nervous because of his rapid pluse. At times, his eyes bulged on the slightest provocation. He was so tired, all he wanted to do was sit or lie down, even after a night's sleep. His physician suggested he energize his thyroid. He was told to enjoy the *Thyroid Nourishing Tonic* every day or until he noted better health. Oscar Q. made it one morning. By mid-day, he was so energized, he felt like an athlete again. His skin was firm, he had an even temperament, regular pulse, normal eye position. Thanks to this simple tonic, he went from "aged" to "athlete" in just one day.

Benefits: The raw enzymes in the tomato-celery juice seize the iodine in the kelp (a high concentrated source of this important gland-feeding mineral) and propel it upwards via the bloodstream to your thyroid, often in moments. *Important:* You-need strong enzymes in order to dispatch the iodine to your thyroid. That is why the kelp is more effective if taken in a raw vegetable juice which brims with these natural activating forces. This combo can create almost immediate thyroid activation. More essential is that this combo creates a balanced and even absorption of the iodine so your thyroid releases a healthy supply of the important hormone.

Adrenal Glands

Location: Shaped somewhat like Brazil nuts, the adrenals are a pair of glands that sit astride each kidney. The adrenals consist of two parts, the *medulla* (central portion) and the *cortex* (covering.) The medulla is partly an extension of your nervous system. It is your body's "alarm box" that prepares you to meet challenges and even danger.

Hormone Function: Your medulla is stimulated by enzymes to produce adrenalin (or epinephrine) which is known as the emergency hormone. When you face stress or danger, enzymes are needed to activate your medulla to release more adrenalin into your bloodstream. Here enzymes use adrenalin to quicken your heart-beat, increase energy-producing sugar in your bloodstream, slow up digestion, fill big muscles with your blood, dilate your eye pupils; in extreme stress situations, enzymes may cause the adrenalin to stand your hair on end.

Your adrenal cortex releases the cortisone hormone which helps regulate metabolism of potassium and other minerals as well as protein, carbohydrates and fat. If you have a weak enzyme supply, your adrenal glands cannot release enough cortisone and related hormones. This may cause blood sugar irregularity, tired-ness, low temperature. *Serious Problem:* A very weak enzyme supply can cause such a cortisone deficiency that your connective tissues as well as skeletal structure begin to deteriorate and this may lead to arthritis.

Basic Guideline: Have an ample supply of enzymes active-ly available in your system to stimulate your adrenal glands so the medulla and cortex release sufficient amounts of health-maintaining hormones.

Enzyme Control Program. Two nutrients are especially im-portant for the active production of adrenal hormones. These are the B-complex and C vitamins. Your enzymes need these vitamins for the regeneration of the membranous covering and the cell-tissue components of your adrenals. It is noteworthy that these glands store *more* of Vitamin C in your body than just about any other body part. Enzymes deposit these vitamins in your tissues for the collagen repair as well as the energizing of the glands to meet the challenges of daily living. So your adrenals may well thrive and survive on the amount of B-complex and C vitamins available to them.

"ECP" for Your Adrenals. To give youthful *Enzyme Catalyst Power* to your adrenals, boost your intake to two special groups of everyday foods containing these nutrients. *Vitamin B-Complex:* Whole grains such as bread, cereals, wheat germ, bran, brewer's yeast, desiccated liver, whole brown rice, soybeans, green leafy vegetables. *Vitamin C:* Fresh fruits (especially citrus), papaya, strawberries, cantaloupe, tomato, broccoli, green peppers, raw leafy greens, sweet potatoes.

"ECP" for Youthful Vitality. Try any of these combinations: (1) To a glass of orange and/or grapefruit juice, add one-half tea-spoon brewer's yeast, one-half teaspoon each of wheat germ and bran. Blenderize and drink slowly. (2) Prepare a fresh fruit or vege-table salad, sprinkle with brewer's yeast and wheat germ. (3) Have whole grain bread or muffins or rolls that have wheat germ added. (4) To a glass of tomato juice, add one-half teaspoon brewer's yeast, one teaspoon desiccated liver (from health store), one-half teaspoon bran. Blenderize. Drink every other day. These are Enzyme Catalyst Power foods and tonics that will so nourish the millions of cells and tissues of your adrenals that a fresh release of balanced hormones will give you sustained energy and youthful vitality, often within minutes.

Enzyme Pep Program for Activated Adrenals

Troubled with poor digestion as well as constant nervous tension, not to mention unrelieved fatigue, Edith McK. thought an exer-cise program would help revitalize her body and mind. It only tired her. Finally she complained to her physical therapist, and she was told that she needed an Enzyme Pep Program to activate her obviously sluggish adrenals. Edith McK. followed the sug-tions for boosting intake of B-complex and C vitamins in every-day food combinations. She used the preceding tonics and salads and also whole grain bread products. Within two days, she en-joyed healthy digestion. An added bonus was such strengthening of her nerves, that she was no longer bothered by petty (or large) annoyances. She had such vigor, she was soon made a part-time instructor in her exercise class! All this was thanks to the speedy effect of enzyme activation of her adrenal glands with the needed vitamins.

Pancreas Gland

Location: Located near your stomach is your pancreas gland. It is a broad strip of soft glandular tissue across the back of your abdomen. It is considered a digestive gland in the sense that its task is to metabolize carbohydrates and send them into the blood-stream in the form of energy-producing glucose.

Hormone Function: Scattered through your pancreas gland are small clusters of specialized cells known as the *islets of Langer-hans.* They produce *insulin,* a hormone released directly into your bloodstream. Pancreatic substances contain digestive enzymes — catalysts by which complex substances in your food are broken down into particles that can then be absorbed by your body for its building and rebuilding. But these enzymes are even more important to your thyroid because they activate your *islets of Langerhans* to produce insulin, which enables your body to make proper use of starch and sugar.

Problem: An insufficient amount of enzymes or weak enzyme content will cause your pancreas to become sluggish. As a consequence, there is a decrease in the production of insulin. This causes excess accumulation of sugar in the bloodstream and runs the risk of development of diabetes. For protection, the availability of enzymes is of vital importance.

Enzyme Control Program. To invigorate your pancreas, the enzyme program calls for increased amounts of raw foods. In so doing, you will be supplying your pancreas with the needed enzymes for insulin manufacture.

Threat of Cooked Food. Does it matter if fruits and vegetables are cooked or raw? Yes, insofar as enzymatic health is concerned. Cooking lowers vitamin content, weakens and destroys essential enzymes. A meal of cooked and processed foods requires from 80 to 100 hours for full digestion. While cooked food is languishing in your intestines, it gives off chemical byproducts that tend to poison some of the systems involved in carbohydrate metabolism. This predisposes to a diabetic condition. The enzyme-starved pancreas cannot resist the onslaught of toxic wastes stemming from smouldering cooked foods and it grows weaker and weaker. Now insulin production declines. Sugar builds up. Diabetes may be the consequence. So for the sake of your pancreas, you could say that cooked food is a threat to your hormone health.

Benefits of Raw Food. Food eaten fresh and raw is high in natural nutrients and the precious enzymes. A meal of fresh and raw food requires from 18 to 24 hours for full digestion. There is a marked reduction in the amount of toxic wastes. The pancreas is now nourished as the food enzymes use nutrients for its building and functioning. There is a more healthful release of insulin because the absorption of carbohydrates from raw foods is smoother and more effective in the digestive system.

Chewing Is Good for Your Glands. Raw foods require much more chewing than do cooked foods. The chewing process alerts various digestive enzymes to perform their task more efficiently in a speedier amount of time. This is believed to be most beneficial for all of your glands which are then actually bathed with regenerating enzymes — you can chew your way to more youthful health.

From "Sugar Overload" to "Healthy Insulin" in Two Days

Unrelieved tiredness as well as constant trips to the bathroom prompted Louis A. to be examined by an endocrinologist (specialist in glands and hormones). He was told that he had a sluggish pancreas. A drop in insulin meant "sugar overload" that threatened to erupt in diabetes. Louis A. was told to follow a simple program: On *alternate* days, eat raw foods exclusively. It was as simple as that. Louis A. followed it. By the end of the second day, an examination showed normal blood sugar levels and a healthy pancreas. The raw food had supercharged his glands with powerful enzymes that promoted the release of ample sugar-dissolving insulin. To avoid "sugar overload" and enjoy "healthy insulin," Louis A. has an even simpler maintenance program: One day a week, totally raw foods! He now feels energetic, makes no excessive bathroom trips and has a youghful insulin-producing pancreas. He no longer faces the threat of diabetes.

Raw foods are catalysts to your pancreas in that they send in stimulating enzymes that activate this sluggish gland and promote the release of essential hormones.

Put power into your pancreas with raw food enzymes.

The Ovaries

Location: In women only, these are situated, one at each side of the uterus or womb, immediately below the opening of the Fallopian tube, just before your abdomen. During fertile years, the ovaries are involved in pregnancy. In later years, they decline in function and may cause the onset of so-called "change of life."

Hormone Function: Healthy ovaries release estrogen, a female sex hormone that promotes fertility and also performs the important function of providing female characteristics. Skin, hair, figure and general feminine qualities are influenced by adequate amounts of estrogen release.

Keeping Youthful in Later Years. When the female body passes the middle years, there is a decline in production of estrogen because pregnancy is no longer desirable. With this hormone decline, there may be subsequent losses of youthful feminine qualities. This can be a problem, but with the availability of enzymes, it can be balanced and corrected.

Enzyme Control Program. Fresh raw foods, especially seeds and nuts are prime sources of polyunsaturated fatty acids. These are health-building substances that are needed by your enzymes to keep the dormant ovaries in healthy condition. In effect, enzymes will metabolize these fatty acids and provide a *substitute estrogen* that can provide feminine qualities similar to the natural estrogen hormone. The aim here is to eat an ample amount of raw seeds and (shelled, of course) nuts. You need to *chew well* so that salivary enzymes can then take out the important ingredients and use them to nourish your slowed-up ovaries. In so doing, the release of *substitute estrogen* can continue and you can hope for extended youth ...at any age.

The Prostate

Location: In men only, the prostate is a rounded mass of muscle fibers and small glands immediately below the urinary bladder, surrounding the outlet of the bladder (urethra). It releases a part of the seminal or reproductive fluid in which sperms from the testes are suspended.

Hormone Function: The prostate releases fluids that are combined with seminal secretions and thereby give fertility to the male. In some situations, an enzyme deficiency may cause the prostate to become weak and this can cause enlargement and related difficulties known as prostatitis.

Enzyme Control Program. Basically, your prostate has to release hormones that nourish your sperm cells. If there is enzyme weakness, then your prostate is unable to manufacture sufficient amounts of such hormones and this causes disorders. *Problems:* In an enlarged and/or inflamed prostate gland, there is a continual urge to pass urine that is painful to release, too. There may be burning sensations. Also, there is a sensation of fullness in the rectum which can be traced to the swollen prostate gland. If this condition is not corrected, it may so infect the prostate that its removal via surgery is required.

Prostate-Pleasing Enzyme Programs. Substances in oils are found to be extremely beneficial to the prostate gland. In particular, the polyunsaturated fatty acids found in egg, rice bran oil, soybean oil, sunflower seed oil, corn oil, peanut oil, wheat germ oil, are needed by your prostate. These are taken up by enzymes and dispatched to nourish the components and segments of your prostate to keep it lubricated. *Suggestion:* Use these oils regularly in your daily food program for prostate-pleasing enzyme reaction.

Enzymes + Zinc = Healthy Prostate. As a trace mineral, zinc appears to be needed by enzymes to help rebuild and regenerate the prostate gland. Enzymes require zinc to create a cleansing of wastes and then the cell-tissue regeneration of the prostate. *Suggestion:* Boost your daily intake of these high-zinc foods: brewer's yeast, beans, nuts, seeds, wheat germ, fish, meat (especially liver). *Guideline:* Eat everything raw, except the fish and meat. In so doing, you'll be sending a powerhouse of enzymes and zinc to your prostate gland and helping to protect it against "aging" degeneration and risky prostatitis.

Zinc-carrying enzymes appear to offer your prostate resistance against weakening and enlargement as you grow older.

Total Body Health Begins with Improved Glandular Health

To enjoy total body health, use enzymes to improve and "wake up" your glands. You will then have "happy hormones" that will give you the look and feel of complete youth...at any age.

IN A NUTSHELL:

1. Enzymes influence the power of hormone production by all of your glands.
2. Note checklist of important glands and how to use enzymes to improve their function.
3. Helen S. bounced back to youthful health on a simple Happy Hormone Program.
4. Oscar Q. went from "aged" to "athlete" in just one day by using a special Thyroid Nourishing Tonic.
5. Edith McK. discovered that an easy Enzyme Pep Program could activate her adrenals and give her super-digestion and amazing youthful vim and vigor...within two days!
6. Louis A. used simple enzyme programs to put new power into his pancreas. He went from insulin-deficient "sugar overload" to enzyme-boosted "healthy insulin" in just two days.
7. Women and men in middle years can use enzymes to take up the slack of body slowdown in the manufacture of ovary and prostate gland hormones.

CHAPTER 6

"Tired Shopper"? Refresh-Revive-Rejuvenate
In 30 Minutes With ECP

That "tired feeling" you complain about when faced with midmorning work may be traced to a sluggish enzyme system. If you yawn and blink your eyes in early afternoon and feel exhausted, then the cause may be in "enzyme stagnation." If you find it more and more difficult to go through your daily tasks, then the fault may be in "sleepy enzymes." If you are a "tired shopper," then you need to wake up your sleepy enzymes and bounce back with youthful vigor.

The "ECP" Way to Enzyme Vitality. The *Enzyme Catalyst Program* calls for helping you wake up and alert your enzymes to meet your responsibilities of daily living. You may be a housewife, a shopper, a commuter, a working person (or a combination of all) and you need daylong (and nightlong) energy to fulfill your obligations. You can do this by having *alert enzymes* that keep active in metabolizing nutrients and using them to give you healthy vim and vigor. The *Enzyme Catalyst Program* will show you how simple fun-to-do exercises can refresh your enzymes and permit them to nourish your trillions of body cells and tissues and actually "drench" them with vitality-producing nutrients.

Example: You have finished a particularly heavy meal. You feel groggy. You lie down and fall into a heavy sleep for an indefinite amount of time. When you awaken, you still feel heavy and tired. *Reason:* The inactivity, especially after a meal, causes your digestive enzymes to become sluggish and even *stagnant*. They cannot work as effectively in digesting and assimilating foodstuffs.

83

Therefore, you are denied full energizing benefits from the ingested foods. *Suggested Solution:* After any meal, go out for a walk. Keep yourself active. It is this circulation boost that tends to put power into your digestive enzymes, enabling them to work more vigorously. They are then able to help distribute energizing nutrients throughout your body...often, in minutes. The trick here is to *activate* your enzymes so they, in turn, will energize your body and mind.

HOW "ECP" EXERCISES GIVE YOU SPEEDY VIM AND VIGOR

The following *Enzyme Catalyst Program* exercises (they take less than 30 minutes) offer you speedy vim and vigor because:

1. ECP exercises help loosen up your tight body spots.
2. ECP exercises help firm up your muscle tone.
3. ECP exercises give you improved reaction capacity and total coordination.
4. ECP exercises improve the flexibility of your joints.
5. ECP exercises boost your circulation.
6. ECP exercises provide nutrient-carrying oxygen to all of your vital organs.

By following these programs just about anywhere, you put new life into your enzyme system...and new life into your body. Most beneficial — it happens within 30 minutes...or less!

SEVEN "SIT-DOWN" EXERCISES FOR
SPEEDY ENZYME REGENERATION

Here is a set of 7 "sit-down" exercises (that's right — you can do them when sitting down anywhere) that will supercharge your enzymes in a matter of moments. You'll feel alert and energized and filled with a desire to accomplish more chores.

1. *Half Head Turn.* Turn your head as far to your right as you can. Return head to front. Turn your head as far to your left as possible. Return head to front. Repeat the entire exercise 10 times. *ECP Benefit:* Loosens up "choked" enzymes in your upper spinal column. Helps you move with more youthful flexibility. Takes the "load" of your back.

2. *Head Tilt.* Bend your head forward (with chin against throat) as far as comfortably possible. Bring your head slowly back as far as comfortably possible. Repeat the entire exercise 10 times.

ECP Benefit: Stimulates the muscles of your neck. Releases "captured" or "locked in" enzymes. Now you can move your head and neck with youthful agility.

3. *Reach For The Sky.* Alternately and rhythmically stretch up your arms as if to pick a star from the sky. Repeat exercise 10 times with each arm and shoulder. *ECP Benefit:* Activates enzymes in the muscles of your shoulders. This creates natural vibratory stimulation so that you feel more youthful flexibility all over your body.

4. *Stomach Bends.* Draw your stomach fully in. Drop your trunk forward while lifting the front of your feet very high. Place your feet hard on the floor. Relax your stomach muscles and raise upright slowly again. Repeat 10 times. *ECP Benefit:* Improves enzymatic blood circulation in your digestive tract; especially stimulating to your bowels to permit freer exchange of nutrients. This helps stimulate improved assimilation and distribution of energy-producing nutrients.

5. *Foot Patting.* Sit with elbows on your knees. Bend forward, pressing your upper body down on your knees. Lift your toes with your heels as high as possible. Drop your heels and lift your toes. Repeat from 15 to 20 times. *ECP Benefit:* Sends released enzymes to improve blood circulation to your feet and lower legs. Helps do away with "tired feet" in a matter of moments.

6. *Partial Standing.* Slowly raise upward from sitting position without the help of your hands until you are standing erect. Sit down. Repeat from 10 to 20 times. *ECP Benefit:* Provides enzyme refreshment of your knee and hip extension muscles. This cleansing-nourishing reaction helps invigorate your body so that you feel alert and energetic in a little while.

7. *Cooling Down.* After completing these six exercises, sit fully relaxed. Breathe evenly and gently, counting slowly to 30. Let the air out slowly. Just as slowly take in deep breaths as you count. Repeat until you feel relaxed and comfortable. *ECP Benefit:* The inhalation-exhalation helps to oxygenate your total body enzymes. So invigorated, they now stream throughout your body, nourishing your trillions of cells and tissues and energizing your organs so you feel young all over!

From "Washed Out" to "Full of Life" in 20 Minutes

Muriel F. felt "washed out" in midmorning. No matter how well she slept the night before, she was still exhausted when the day was hardly underway. Complaining to a physical therapist, she was given the preceding set of easy-to-follow enzyme-activating exercises to follow. She did so. Muriel F. experienced such revitalization within 20 minutes, that she felt no longer "washed out" but "full of life." Now she could carry out her daily and even nighttime obligations with the vitality of a youngster. Her enzyme system had become supercharged with this program. The enzymes invigorated and regenerated her various systems and gave her youthful vitality.

"ECP" EXERCISE PROGRAMS WHILE WAITING ON LINE

Sometimes it seems as if you're always waiting. In traffic, for gas, at the market, in the bank. Make good use of this time to energize your body. You can follow *Enzyme Catalyst Programs* while you wait just about anywhere, seated or standing. You'll find yourself becoming supercharged with exhilarating energy and the waiting will hardly be felt.

Basic ECP Benefits: The following exercises are designed to "release" and "distribute" the "locked in" enzymes throughout your body. Once dispatched, these biological wonder workers can nourish and revive your sluggish cells. They also help to nourish your vital organs and thereby give you a refreshing new appearance and attitude. They take only moments but can provide you with day-long energy, and they work almost immediately.

Exercise in Your Car

(Remember: *NEVER* exercise in a moving car. Perform *only* when sitting and waiting in your car.)

1. *Bumper To Bumper.* Hold your steering wheel. Tighten muscles in your buttock. Lift right cheek off the seat. Lower it to the seat; lift left cheek. Pull in stomach and keeping buttock muscles tight, lift and lower each cheek five times. *ECP Benefit:* lessens driver fatigue.

2. *Rubber Necker.* Hold your steering wheel. Raise shoulders toward your ears; lower them. Raise and lower them five times. Now, move them in the following circular motion: lift toward ears, push forward, then down, then as far back as comfortably possible. Repeat forward circular motion five times. Reverse action for five backward circular motions.

Next: Relax shoulders, letting your head fall forward. Move it in a counterclockwise direction, going from chest, to left shoulder, to back, to right shoulder, to original forward position. Reverse direction of head motion to clockwise direction. Repeat exercise 10 times. *ECP Benefit:* Relaxes your neck muscles.

3. *Spare Tire.* Hold your steering wheel. Tighten all abdominal muscles. Hold tight five seconds, then relax. Repeat exercise 10 times. *Tip:* To make it more effective...twist torso to left and then to right, while keeping abdominals tight. Do a total of 10 twists to right; 10 to left. *ECP Benefit:* Firms and tones stomach; helps remove "spare tire" through enzymatic action upon stored up fat in this region.

4. *Fender Bender.* (1) With car parked well off the road, place right hand on left front fender. Using right hand for balance, lift left leg, bending it at knee. (2) Clasp left ankle with left hand and pull on ankle, raising left foot over buttocks. Hold position for five seconds and release — without lowering foot to ground. Repeat exercise 8 times on each leg. *ECP Benefit:* Stretches fronts of thighs, tightens and tones backs of your thighs and buttocks.

Exercises While in the Store

1. *Turning Cart Handles.* Clasp bar of shopping cart (the kind that turns) with both hands palms down. Turn bar forward with hands. Repeat turning movement 10 times. *ECP Benefit:* Stretches forearm muscles; releases tension in forearms and shoulders.

2. *Checking Out.* (1) Turn chin over right shoulder. Slowly turn it, as far as it is comfortable to your left shoulder. Turn 3 times to your right; repeat to your left. *ECP Benefit:* Releases neck tension. (2) Keep hips still and twist torso. First to right, then to left. Let your head and neck move with your torso. Twist 6 times to your right, then 6 times to your left. *ECP Benefit:* Releases back tension.

3. *Bend For The Goods.* Holding shopping cart, lower into a slight knee bend, tightening abdominal muscles. Repeat 8 times. *ECP Benefit:* Improves health of your stomach and legs.

4. *Bored Feet.* Holding your shopping cart for balance, lift onto the ball of your right foot. As you return right heel to the floor, lift onto the ball of your left foot. Pull in your stomach and tighten your buttocks. Repeat foot alternations 10 times. *ECP Benefit:* Enzymes help wash away lactic acid deposits that cause fatigue; the enzymes then help stimulate sluggish circulation so that you have more vitality in your feet...and your entire body, too.

Suggestions: Follow these *Enzyme Catalyst Program*

exercises whenever you fell the need for a lift. In a matter of moments, enzymes become freed from their congestion. They now (1) wash away fatigue-causing waste deposits and (2) send a stream of energy-producing nutrition flowing throughout your body and mind. You'll discover healthful energy and a new zest for living.

HOW TO TAP YOUR HIDDEN SOURCES OF "INSTANT ENERGY"

Prolonged inactivity creates a problem. Your enzymes go on a "sit-down strike." They become sluggish. They lose much of their catalyst powers. Now you start to feel a bit dazed. You're not as steady on your feet as usual. Your back feels as if you've been lugging heavy sacks for hours. *Reason?* Your choked up enzymes are trapped in debris. They cannot perform their never-ending work of cleansing and rebuilding your entire body. *What To Do?* Simple. You need to tap your hidden sources of "instant energy." You do this through a series of effective enzyme stimulation exercises. Some take a few minutes. They work just as speedily, too.

Suggestions: Follow as many of these programs throughout the day as possible. They are all aimed at breaking up accumulated lactic acid and related waste products and freeing enzymes so they can provide your needed "instant energy." The more of these easy and fun-to-do exercises you follow, the more refreshed and rejuvenated your body will become. Now...go to it.

*Walk...walk...walk! Park your car a few blocks from your destination and walk the rest of the way. When possible, plan to walk whenever you can after each of your regular meals. This boosts the power of enzyme action.

*Use your feet instead of the elevator. Take to the stairwells, slowly, but regularly. Stair climbing helps activate your enzymes and boost their waste-burning powers. The enzymes will then increase the oxygen nourishment of your heart, lungs and respiratory system. You'll feel healthier and more vigorous in a few moments.

*Stand as much as comfortably possible. An upright position tends to keep enzymes more active. More enzymes are energized if you walk around the room instead of constant sitting.

*If you walk (ordinary walking) just 20 minutes a day, you'll supercharge your oxygenated system with vitality that will "wake up" your enzymes and help them cleanse and regenerate your internal organs. Regular walking is an effective way to "exercise" your enzymes and boost basic health.

*Keep moving while you talk on the phone. Stand up. Move and walk around as far as the line cord will permit.

*Keep an old tennis ball around. Frequently, squeeze firmly with each hand. Tends to "milk out" lactic acid accumulations inside you, and free "choked" enzymes so they can work more efficiently in rebuilding body vitality.

*Try moving walls: stand one pace away from a wall. Place your hands against the wall at shoulder height. Press forcefully as if you could move the wall. Count to ten and relax. Do this often to create rhythmic exercise while standing in one position. Enzymes become energized as the wastes are sloughed off and rebuilding occurs. You'll feel refreshed in a matter of moments.

HOW TO EXERCISE YOUR ENZYMES AND SELF-REJUVENATE IN TEN MINUTES

Become a go-go whirl! Switch on a record or tape with a frantic dance theme. Exercise for 10 minutes in a go-go style: rocking and rolling, twisting and turning, reaching and stretching.

The more imaginative you are, the more exercise is given to your sluggish enzymes. Your entire body becomes liberated from congestants. Enzymes stream through your circulatory system to refresh every part of yourself. Let the music keep you go-go-going without interruption. It's the "fun" way to exercise your enzymes and rejuvenate yourself. Just 10 minutes. You'll feel day-long enzyme refreshment.

"E-E" Break for Speedy Refreshment

Before Barbara H. has to tackle any task, she refreshes herself with a 10-minute "Enzyme-Exercise" Break. She does the preceding go-go motions in her room. She was given this suggestion by her physical fitness instructor. She is able to enjoy speedy refreshment as enzymes actually "make her over" because of this simple program. She calls it her "E-E" break that re-makes her from head to toe. All this for just 10 minutes of motion!

15-Minute Walk After Any Major Meal for Enzyme Enrichment

Wherever and whenever possible, take a walk after a major meal. *Benefit:* Enzymes may become overburdened with the onrush of food and may cause digestion to be sluggish. This reduces the efficiency of cellular regeneration. But with a simple walk, the gentle rhythm promotes a natural "vibratory" stimulus upon the digestive tract. Your enzymes are stimulated and able to work more efficiently for food assimilation. A simple 15-minute walk can work wonders of self-rejuvenation through enzyme stimulation.

QUICK ENZYME "PICK UP" TONIC

> 1 cup of orange juice
> 1 teaspoon of brewer's yeast
> 2 tablespoons of protein-fortified powdered skim milk
> 1 egg

Blend all ingredients for 20 seconds. Drink slowly.

Benefit: Raw enzymes in the orange juice take up its cell-tissue building Vitamin C, combine it with the high protein and B-complex vitamins of the other ingredients to create "food" for your digestive and body enzymes. Within minutes, a surge of vitality occurs so that you can refresh-revive-rejuvenate and enjoy all of the tasks ahead.

Suggestion: Make this *Quick Enzyme "Pick Up" Tonic* in advance. Put it in a thermos. Carry it with you. Whenever you feel the need for refreshment, just take a small portion. Your body enzymes will become supercharged with healing power. You will feel alert and alive...no matter where you are!

ENZYME-ENERGY FOODS YOU CAN CARRY WITH YOU

An important part of your *Enzyme Catalyst Program* does consist of fresh raw foods and drinks as well as self-cleansing exercise programs. A connecting link to enzyme invigoration is the use of several high-energy foods. You can carry these with you. A handful of them will tend to "wake up" your "tired enzymes" and supercharge you with youthful vigor. Carry any or all of these in a small packet or any handy container. Chew thoroughly before swallowing. Then get ready for renewed vitality...no matter where you are.

Raisins. High concentrations of natural fruit sugars. The raw enzymes tend to catalyze accumulated wastes in your system, free body enzymes and set off a chain reaction of speedy revitalization. Munch them regularly. Be sure to use the sun-dried variety. Avoid those that have been chemically dried since residues of this contaminant can be destructive to enzymes.

Figs. Locked within the microscopic seeds are powerhouses of enzymes. They work by immediately elevating your blood sugar levels, washing away tiredness, helping you feel that "get up and go" reaction, very speedily. Sun-dried or fresh figs are great sources of this enzyme energy powerhouse. You might eat them as part of your breakfast for long, long hours of vigor.

Sun-Dried Fruits. Great source of energy-producing enzymes. When fruit is carefully dried under the sun, plant physiology causes its enzymes to become highly concentrated. Just a few pieces of any desired sun-dried fruits will have "dynamite" vitality effects …within moments.

Banana. Encased in soft, nonirritating banana cells are powerful energy-creating enzymes. These unite with this fruit's complex carbohydrates and create easily digested natural sugars. Invert sugar in the banana is used by enzymes to give you speedy metabolism. Just one banana and you feel that you're the top banana of the bunch in terms of enzyme-created vitality!

Orange. Here is a prime source of Vitamin C which is speedily used by enzymes to promote cell-tissue growth and regeneration of your body parts. Eat an orange or two and be rewarded with enzyme power in a few moments.

Seeds. Sunflower, squash and pumpkin seeds are high concentrations of polyunsaturated fatty acids as well as living enzymes that work to wash and scrub and then energize your own body catalysts. They are rich in plant protein; your enzymes take this protein and break it down into almost perfect amino acid balance that helps regenerate your own vital organs. Munch a handful of seeds whenever you want to experience more vitality.

Nuts. Preferably the unsalted and fresh variety; you may use dry roasted but avoid any that have been heated in fats. They are difficult to digest and enzymes are depleted or weakened by this high heat. Chew a handful of nuts and be rewarded with an internal cleansing as the plant protein and complex carbohydrates work with minerals to actually nourish your enzymes and give them super power. Whenever you're tired, chew some nuts. You'll get an "instant lift" that will make you feel youthfully alive.

KEEPING YOUR ENZYMES ALIVE

Keep your body active with movement and these enzyme-alerting foods, and you should be able to say goodbye to fatigue.

Take advantage of every opportunity, no matter how small, to activate your enzymes. Your day is filled with hundreds of opportunities to keep moving. Take the stairs, rarely the elevator, unless it is a very steep climb. Don't use your auto or a bus for short trips. Try not to remain seated for too long. Use your lunch break to move about.

Before settling down to watch television at home or read a book, take an evening walk.

Whenever possible, walk or pace around if you have to do much talking.

Basic Program: Put aside a regular part of your daily schedule for this type of enzyme exercise. *Once Doesn't Count.* It is not the occasional effort but regularity and stick-to-it habit that will boost enzymatic health. Start slowly, but keep it up.

Follow the nutritional programs suggested to further activate your clogged up enzymes. Methodically move your joints through their full range at least once a day. *Example:* tense and flex. Or, reach and stretch. Do this regularly and your liberated enzymes can do a major "body housecleaning" that will make you as fresh and alive as brand new and youthful Springtime.

Are you always complaining about feeling so tired? Does the day drag on and on? Is life passing you by because you can't physically and emotionally partake of all the joys available for the taking? Then wake up your sleepy enzymes. Try the *Enzyme Catalyst Program* as outlined in this chapter. You will be able to enjoy "total refreshment" within 30 minutes or less. Now you'll be able to wake up and live a life of healthy joy...at any age!

MAIN POINTS:

1. *Enzyme Catalyst Programs* ("ECP") offer you a set of enzyme-rejuvenating reactions within minutes.

2. Follow the set of seven "sit down" exercises that regenerate your enzymes (and yourself) in minutes. Can be done anywhere.

3. Muriel F. went from "washed out" to "full of life" in 20 minutes on the set of seven exercises.

4. You can do enzyme exercises while sitting and waiting in your auto!

5. Refresh yourself with enzyme exercises while in your local supermarket!

6. Release enzymes for "instant energy" with any of the described simple programs.

7. Barbara H. enjoys speedy refreshment with a fun-to-do "E-E" break.

8. Prepare the *Quick Enzyme "Pick Up" Tonic* and be supercharged with vitality almost at once.

9. Enjoy the assortment of chewy good enzyme-stimulating foods to be used almost anywhere.

CHAPTER 7

The Enzyme-Collagen Way to

Youthful Digestion

Collagen is described as the "missing link" for self-rejuvenation.[1] The presence of collagen in your digestive tract can determine the power (or weakness) of youthful digestion.

Let's look more closely at this "missing link" in enzyme health and see how to boost its production so you can enjoy better health through more youthful digestion.

What Is Collagen? A protein-like substance made within your body. It serves as a cement to hold together the trillions of body cells and tissues.

How Does It Boost Health? It is needed to strengthen and rejuvenate all the cells and tissues in your body. Enzymes use collagen to nourish and heal bones, body organs, blood vessels and glands. Enzymes use collagen to stimulate natural cell growth and carry on the process of the never-ending cycle of internal digestion and healing.

What Are Deficiency Risks? Without it, wounds do not heal; cells break down; tissues disintegrate; skin sags. There is a breakdown of the health of your bones, organs, vascular components and glands. Furthermore, vitamins and protein cannot be fully metabolized and illness may occur.

How Can Collagen Be Produced Within Your Body? With the intake of enzyme-containing foods as well as Vitamin C, zinc

---

[1]*Parker Natural Health Bulletin,* West Nyack, New York 10994. Vol. 9, No. 22. October 22, 1979. Available by subscription.

and other elements. Enzymes now can bring about important collagen production.

How Will Enzymes Boost Collagen Formation? Found solely in seeds, nuts, raw foods and fresh fruit or vegetable juices, enzymes are needed to create a catalyst reaction. That is, enzymes will take raw materials from your body and bring about the formation of collagen. Now your trillions of body cells and tissues will become strengthened and rejuvenated. *Tip:* Eat fresh fruits and vegetables (and drink their juices) regularly. Eat healthful seeds and nuts for high concentrations of enzymes.

How Will Vitamin C Increase Collagen Supplies? Enzymes use Vitamin C to manufacture collagen between your cells; there is reinforcement of other intercellular material, similar to the way that steel rods reinforce poured concrete. Good Vitamin C collagen-making food sources are — oranges, grapefruits, tangerines, lemons, limes, papaya, strawberries, cantaloupe, tomatoes, broccoli, green peppers, raw leafy greens and potatoes.

How Is Zinc Helpful In Collagen Formation? During ordinary wear and tear, much collagen is broken down in your body. Enzymes require zinc to invigorate the DNA and RNA components which control cellular reproduction. Enzymes use zinc, with the help of collagen, to stimulate the manufacture of DNA and RNA to help form needed new cells and tissues. Without zinc, your enzymes are weak and cannot create healthy DNA-RNA segments. Results? Collagen manufacture slows down and the aging process follows. Try these zinc-collagen making food sources — lean meat, liver, eggs, seafood, skim milk and whole grain breads and cereals.

Collagen may well be the "missing link" required by enzymes to help influence body rejuvenation through cellular proliferation. An ample amount of food enzymes will bring about improved collagen rejuvenation of your vital organs.

DAILY INTAKE OF RAW FOODS IMPORTANT FOR ENZYME DIGESTIVE HEALTH

Basic Program: Each meal should *begin* with a raw vegetable salad or any seasonable assortment of fresh raw vegetables. Use desired dressing. Each meal should *end* with a raw fruit salad or seasonal assortment of fresh raw fruits. *Basic Benefit:* You send a power-packed supply of enzymes shooting throughout your system so that ingested foods can be better metabolized and collagen be formed to heal and bind the components of your digestive tract.

Youthful Health Begins with Healthy Digestion. The foods you eat may be extremely nourishing but if their vital elements are not enzyme-digested, you will run the risk of illness. Furthermore, a healthy digestion begins with powerful collagen-creating enzymes. This emphasizes the value of fresh raw foods as frequently as possible. You need a plentiful supply of these enzymes in order to enjoy healthy digestion and youthful health.

How Enzymes Can Rebuild Total Health

Through *digestion,* your enzymes are able to create collagen and rebuild total and youthful health. Briefly, let's understand the powers of this miracle worker.

An enzyme is a digestive catalyst. It produces a reaction but does not combine with the end product.

Rebuilding Process. Enzymes reduce complex food substances to simpler compounds that are acceptable to your bloodstream. Enzymes use these compounds (such as collagen) to rebuild your trillions of body cells. Enzymes use the same collagen to strengthen your digestive organs so that they can continue their function of keeping you nourished.

Enzymes Have Individual Functions. Each enzyme is specific in its action. That is, it acts only upon one class of food substance. *Examples:* the enzymes that act upon carbohydrates do not act upon protein or fats or other nutrients. Also, during digestion, the enzyme that acts upon maltose sugar is not capable of acting upon lactose sugar. So you see that since each enzyme has a specific function, the weakness or absence of just one can cause body distress.

For improved and energetic health, you need an available supply of all groups of enzymes.

Better Digestion — Better Health

Improved digestion will cause improved formation of collagen throughout your body. So you begin with thorough chewing.

Digestion commences in your mouth. The process of chewing breaks up foods into smaller particles. A saliva enzyme, *ptyalin,* acts upon starch, breaking it down into a complex sugar (maltose) and making it ready for further assimilation. Other mouth enzymes work upon different foods to prepare them for your digestive-assimilative tracts. So *chew* all foods thoroughly for better digestive and body health.

HOW TO HEAL DIGESTIVE DISTRESS WITH ENZYME-COLLAGEN PROGRAMS

To Begin: Good digestive health requires enzymes. They are needed to break down protein, carbohydrate, fat, etc., from foods. A deficiency means these nutrients are improperly metabolized, collagen formation weakens and now digestive distress occurs. Enzymes hold the key to digestive and body health.

Here are typical digestive disturbances and their enzyme catalyst correction remedies.

INDIGESTION. Also known as dyspepsia, it is your body's outcry at enzyme deficiency. *Enzyme Correction Remedy:* Emphasize intake of all green and yellow fruits and vegetables *before* you eat a meal. *Benefit:* Enzymes will take the high concentrations of Vitamin A (in the form of carotene) and Vitamin C and use them to create important collagen that will help knit and bind the worn organs of digestion. You'll find more digestive comfort and better health in a little while.

HEARTBURN. Water-diluted enzymes are unable to function and collagen-weakened cells of your esophagus (food tube) are irritated at the approach of foods. *Enzyme Correction Remedy:* Do NOT drink any liquids for 30 minutes before a meal or 30 minutes after a meal. *Benefit:* You will save your enzymes from dilution (drowning) and they will be able to perform their task of digestion and collagen building.

NAUSEA. Ingested food cannot be properly (if at all) metabolized because of severe enzyme deficiency. Food tends to "return" because it is "rejected." *Enzyme Correction Remedy:* Prepare your digestive tract with an *energy enzyme* in the form of fresh fruit in season. Chew thoroughly. *Benefit:* You set off the enzyme action so that ingested food will then be better metabolized. It's important to *begin* your meal with this high energy enzyme fruit such as an orange, grapes, banana, pear, seasonal melon, plums, pitted cherries.

GASTRITIS. An inflammation of the mucous lining of the stomach; collagen is needed to repair this tract so that it will not be raw and enflamed. *Enzyme Correction Remedy:* Prepare fresh vegetable juices of any seasonal plants. Drink single juices or any favorite combo. For one day, drink nothing else but these vegetable juices. Try to avoid any solid foods. *Benefit:* Raw vegetable juices are highly alkaline; enzymes will use this substance to neutralize and buffer the burning sensation and also to soothe the burning of irritated gastric cells while rebuilding the protective collagen coating.

COLITIS. A disorder in which there may be recurring bouts of constipation and then diarrhea. One underlying reason here is that the collagen sheath on the intestinal tube has become eroded and its muscular contractions made weak because of inadequate enzyme regeneration. *Enzyme Correction Remedy:* For one to two days, go on a raw vegetable fasting program. You may have any seasonal vegetables you desire. Avoid dressings except oil which is mild and soothing. Also enjoy freshly prepared vegetable juices. Eliminate salt or sugar or volatile seasonings. *Benefit:* Without interference of other foods, enzymes in the raw vegetables and juices can work "full force" in creating collagen that rebuilds the weakened and damaged colon muscles. Within two days, healing should be enjoyed.

DIVERTICULITIS. Inflammation of the colon pockets, obstructed by swelling or impaction of wastes. The bacterial infection may subside in a day or so, or continue on to cause abscess, spasm or even obstruction of the colon. *Enzyme Correction Remedy:* Boost intake of fiber. This helps to create an increased transit time of wastes and a cleansing action, too. *Benefits:* Fiber is broken down by enzymes and its components used for the making of collagen that then knits and binds the colon and creates strength and vigor. Good fiber sources are carrots, celery, nuts. The most concentrated form is pure unprocessed *bran,* available at most health stores and supermarkets.

Heals Digestive Upset in Three Days

Troubled with colitis, James I., a travelling salesman, made frequent bathroom trips. This was an inconvenience, especially when travelling on the road. He developed digestive unrest that was considered diverticulitis. He had recurring abdominal cramps (when the muscle went into spasm) and lower pelvic aches. Medicines made him react with side effects. When he described his trouble to a registered nurse to whom he sold products, he was given a basic enzyme-collagen healing program. He began to eat four tablespoons of bran with his cereal at breakfast; another four tablespoons of bran with a luncheon soup or stew; a final four tablespoons of bran as part of his evening main entree. Bran was *not* to be taken by itself because it was too irritating. Simple program. Results? In three days, James I., said "goodbye and good riddance" to his digestive upset, including colitis and diverticulitis. He had enzyme-collagen-renewed organs and felt reborn. Indeed, he was "remade" thanks to the regenerative powers of enzymes.

FREEDOM FROM CONSTIPATION WITH ENZYME-COLLAGEN

Problem: The intestines fail to respond as often as they should. Weakened muscles cannot perform *peristalsis* (wave of muscular contraction needed for rhythmic movements) and waste matter is accumulated. Constipation may be habitual or frequent enough to cause discomfort. Toxic wastes back up into the bloodstream and cause destruction of important substances. If constipation is neglected, the body starts to decline.

Enzyme-Collagen Solution: Before your breakfast, eat several freshly washed whole apples. The more tart the variety, the stronger the enzyme-collagen content. Ten minutes later, follow with a glass or two of room temperature water.

Benefit: Enzymes will seize upon the apple's pectin and natural fruit sugars and energize the formation of collagen. Furthermore, healthy apple acids create a strong reaction upon accumulated wastes. They help boost the power of the intestinal tract and promote a normal movement.

"Natural Laxative" Ends Lifelong Constipation Problem

Peggy R. suffered from constipation as long as she could remember. Laxatives weakened and also eroded her collagen covering so her intestinal tract could not function under its own natural power. A herbal pharmacist told her of this simple remedy. Just chew several freshly washed apples and then drink room temperature water. Within moments, Peggy R. experienced relief. She continued this program for the rest of the week. Soon, she could forget her lifelong constipation problem. It was solved by enzyme-collagen regeneration of her once-weakened intestinal tract.

HOW TO USE ENZYMES TO TAME DOWN RUNAWAY BOWELS

Diarrhea. Uncontrollable trips to the bathroom and an excessive amount of waste passage can drain out the body's resources. Nutrients cannot be absorbed and this may cause further malnutrition.

Enzyme Healer: For two days, eat slightly steamed (in oil) vegetables. This preserves much of the enzyme and nutrient content. It's important to select high cellulose (fiber) vegetables such as carrots, turnips, broccoli tops, celery, cabbage, mustard greens, brussels sprouts, cauliflower. Eat singly or in any desired combination.

Benefit: There is a comforting warmth to these enzyme foods when they are slightly steamed. They are also easily digested, considering the collagen depletion problem. These enzyme foods relieve diarrhea by increasing bulk, acting as a "plug" and also *regulating* the transit time in the gastrointestinal tract. Enzymes use collagen to heal this tract and soothe it so that it does not work too rapidly and induce diarrhea. Just two days will help tame down your runaway bowels. Afterwards, you should be able to return to a basic healthful food program. But take out "enzyme insurance" for your intestines by eating large amounts of succulent raw vegetables daily.

HOW TO BOOST ENZYME-COLLAGEN REPAIR OF YOUR BODY

You need to have a stronger digestive power. You can thereby boost enzyme-collagen repair if you protect yourself against any weakening forces. *Examples:*

1. Eat but do not overeat. You will fill yourself with so much food that your enzymes cannot accommodate them and this causes ill digestive and body health.

2. Do not eat if you are too tired. Your digestive system is too weak and your enzymes are sluggish...as is the rest of your body. It is best to rest a while and then partake of your meal.

3. Do not eat if you are anxious, fearful or angry. Eating under these unfavorable circumstances will cause a constriction of your digestive organs and enzymes will become "choked" and thereby incapacitated. Wait until you feel emotionally content before you start to eat.

4. Restrict or eliminate the use of sharp condiments, such as salt, pepper, vinegar, ketchup, mustard, salted sauerkraut, pickle relish as well as any volatile items. These retard digestion, tend to "burn" the enzymes and extinguish their powers. Instead, try more flavorable and flattering herbs and spices.

5. Eat casually and comfortably. Do not eat too fast. This is a sign of anxiety and tension or stress. Your enzymatic system becomes just as tense and cannot properly metabolize foods. You'll pay the penalty in digestive (and body) upset later on. Collagen cannot be formed if food is rapidly gulped down without adequate mouth salivary enzyme digestion. If you can't eat comfortably, delay eating until you are rested.

6. Try not to swallow air with your foods or beverages. Called *aerophagia,* it tends to "bloat" your digestive system and block full enzyme catalyst collagen formation. Air swallowing is also a characteristic of nervous eating. So slow down. Breathe through your nose. Don't swallow air!

7. Chew thoroughly. You'll prepare food for enzyme catalyst action and digestive rebuilding much more effectively.

8. Emphasize more free-flowing liquid fats. They appear to provide essential fatty acids for your enzymes and promote better manufacture of healing collagen.

9. Cut down on your intake of hard fats. Trim off visible fats before cooking meats; trim off fat that remains on cooked meats. Pour off melted fats. Don't use them. These hard fats tend to clog your trillions of cells and tissues and actually create a "coating" that resists collagen formation. They can cause tissue *necrosis* (destruction) and predispose to general body breakdown. Fewer hard fats will mean more healing collagen.

10. Do NOT wash down your foods. Drink liquids about 30 to 60 minutes before a meal and the same amount of time *after* a meal. This helps protect against diluting your enzymes. Let them work without being drowned!

VITAMINS + MINERALS = ENZYME COLLAGEN POWER

Your enzymes depend upon vitamins and minerals. They work together. If you have a deficiency, you run the risk of enzyme dysfunction and this can cause weakness of your digestion, the root of health. So you need a balanced daily intake. Every day, get an adequate amount of these important enzyme helpers in these foods:

Vitamin A. Fish liver oils, dairy products, carrots, cantaloupe, peaches, squash, tomatoes and all yellow and green fruits and vegetables.

Vitamin B-Complex. Brewer's yeast, wheat germ, bran, whole grain breads and cereals, brown rice, organ meats such as liver, poultry, desiccated liver, bananas.

Vitamin C. All citrus fruits, green peppers, many vegetables.

Vitamin D. Egg yolk, fish, fish liver oils, fortified dairy products.

Vitamin E. Most vegetable oils, wheat germ, bran, beef liver, seeds, nuts, soybeans, whole grain cereals and breads.

Minerals. These include calcium, phosphorus, iron, copper, magnesium, manganese, potassium, selenium, zinc and the other trace elements. You'll nourish your enzymes with minerals found in these foods: dairy products, whole grains, green and yellow vegetables, most seafoods, liver, lean meats, brewer's yeast, desiccated liver, egg yolk, seeds, nuts and wheat germ.

Suggestion: When planning your meals, include a variety of these foods every day. You'll be providing your enzymes with valuable co-workers who are able to build and rebuild the collagen lining of your entire body.

HOW TO ENZYME-EAT YOUR WAY TO YOUTHFUL HEALTH

"Nutritionally concerned individuals should try to eat more whole grains, fruits and vegetables but less sugar, fat and salt," says Professor Louise W. Hamilton, a registered dietician and teacher of foods and nutrition at Penn State University. Here are some suggestions offered by Prof. Hamilton to help you enzyme eat your way to youthful health:[2]

Boost Fiber Intake. Sources of fiber include whole grains, fruits and vegetables. Prof. Hamilton says fiber "should be increased in our daily diet plan."

---

[2]*Parker Natural Health Bulletin,* West Nyack, New York 10994. Vol. 8, No. 20. September 25, 1978. Available by subscription.

Iron Is Important. To get enough of this mineral (which works with enzymes for collagen manufacture) "concentrate on meats, praticularly liver, and on whole grains. Dried fruits such as raisins and molasses and egg yolk have a lot of iron," suggests the professor.

Dining Out. "When eating in restaurants, avoid fried foods. These provide too many calories and fat. Soft drinks provide too much sugar. Milk-based beverages are better than soft drinks, but keep within your caloric needs. Salad bars are a good addition to fast food restaurants as well as to other restaurants. It's a good sign, nutritionally speaking," comments Prof. Hamilton.

Preparing Foods. "How food is prepared is important. Fried foods are higher in calories and more difficult to digest. Eat boiled, broiled or baked foods."

Cooking Vegetables. Prof. Hamilton says that overcooking vegetables is an enzyme-destroying mistake. "It destroys nutrients, color, taste, texture. Cook only until succulent, in a minimum of water."

Losing Weight. "Eat less, but choose each food carefully for the nutrients it provides."

Basic Guidelines. Prof. Hamilton says, "You must eat a variety of foods to maintain the balance of nutrients your body needs. Each nutrient has a SPECIFIC USE in your body. These nutrients work together. You can eat your way to better health with this proper enzyme balance."

You *can* enrich your health by enzyme-collagen nourishment of your digestive system. Within hours, your newly repaired organs of digestion can help you look and feel more alive than ever before! It's all up to you...and your selection of foods!

SUMMARY:

1. Boost enzyme production of collagen and have a youthful digestive system.
2. Daily intake of raw foods is a "must" for enzyme digestive health.
3. Check your digestive problem and the enzyme correction remedy.
4. James I. healed his colitis problem in three days with a simple enzyme-collagen food.

5. Peggy R. used a tasty "natural laxative" and ended her lifelong constipation problem within a week.

6. Boost enzyme-collagen repair of your body with the 10-step home guide plan.

7. Be sure to include adequate amounts of vitamins and minerals each day because they work with enzymes for the rebuilding of your health.

8. A professor offers you a simple way to enzyme-eat your way to youthful health.

CHAPTER 8

Enzyme Foods That Wash Your Arteries, Melt Body Fat, Put "Go Power" Into Your Heart

A fat-dissolving substance found in everyday raw foods can help keep your arteries young, wash excess fat out of your body and add energy-packed years of youthful life to your heart. This substance, the *enzyme,* has the power to metabolize and then *split* hard fats so that there is less of a risk of excessive accumulation. But even more important is that enzymes are able to cause a life-saving reaction within your system that may hold the key to having a healthy heart. Enzymes boost production of certain substances that offer protection against fat and cholesterol excesses. Enzymes appear to be the life-saving factor that your body must have in daily supplies. Let's see how fat accumulation can be a risk factor and how enzymes come to the rescue.

FATS: THE GOOD VS. THE BAD

Fats Are Good. Fats are an important part of your diet. They give you essential fatty acids which are needed to maintain basic health. Fats assist your body in assimilating important Vitamins A, D, E and K which are used for building healthy skin, bones, cellular strength and resistance to ailments. A reasonable amount of stored fat will act as a shock absorber in your body. This fat will cushion bones and vital organs, protecting them from impact. In colder weather, this added fat will insulate your body against heat loss. Fat combines with the mineral phosphorus, to form a phospholipid substance that is required to build and rebuild your billions of body cells and tissues from head to toe. So you do need fat!

Fats Can Be Bad. With this understanding, you may wonder why fat has received so much "bad press." The reason here is that some types of food fats you eat can become stored to excess in your body. When this occurs, you have what is known as an "overload." This may lead to problems of excessive cholesterol, as well as clogging of your arteries and the risk of heart trouble. Furthermore, there are "good" fats and "bad" fats. Briefly, here is a checklist.

"Good" Fats. Free-flowing oils that are considered polyunsaturated or containing substances that do not harden.

"Bad" Fats. So-called "hard" fats that are considered saturated and also a source of excessive cholesterol.

Which Fats Are More Desirable? Foods containing fats that are *low* in saturates and *low* in cholesterol (also *low* in calories) appear to be more desirable. They offer you appreciable amounts of fats but not to an excess. At the same time, foods that are *high* in polyunsaturates are considered to be preferable in your quest for controlling fat buildup in your body. (See the tables on pages 109, 110 and 111.)

Cholesterol: Keep It Under Safe Control

Basics: Cholesterol is a fat-like substance that is essential to the health of your body organs. It is produced by your body. But it is also found in foods (exclusively of animal origin). *Problem:* Eating an excess of foods from animal sources can cause what is called a "cholesterol overload." If this happens, the fatty deposits tend to cling to your arteries and predispose to a condition known as atherosclerosis. That is, there is a clogging of blood vessels by excessive cholesterol and other materials and this may lead to a heart attack or stroke. *What To Do:* Begin by controlling your intake of excessive amounts of saturated and cholesterol-containing foods. Then, give your body the working materials found in raw foods that will create a dynamic, almost miraculous reaction that will help dissolve and wash out excessive fatty deposits from your system.

TWO FAT-FIGHTING, LIFE-SAVING FRUITS

Two everyday fruits, the *lemon* and the *lime,* contain high concentrates of acid-based enzymes that are able to pierce the thick clumps of fat in foods you have just eaten, and fats that have accumulated in your body, and help liquefy them and prepare them for elimination.

Basic Guideline: If you are preparing to eat a meal that is heavy in fat, your should peel and slice either one lemon or one lime. Remove the seeds. You may even section or quarter the citrus fruit. As you eat the high-fat meat food, also include a section of the citrus fruit. Be sure to chew thoroughly. This alerts digestive enzymes to the task of dissolving the fat. At the same time, salivary enzymes will help break down the fat in the food while you eat it.

But *most important,* the acid tart enzymes in the *lemon* or *lime* fruit will create a "dynamite" dissolution of much of the fat that you eat at the same time.

Content of Fat in Selected Food*

Meats	Calories Serving	% Calories From Fat	Saturated Fatty Acid Rating #	Cholesterol Rating #
Bacon, 2 slices	90	80	High	Moderate
Bologna, 1 slice, 1 oz.	90	82	High	Moderate
Frankfurter, one, 2 oz.	180	80	High	Moderate
Lambchops, trimmed, 3 oz.	160	43	Moderate	High
Liverwurst, 1 oz.	90	78	High	High
Ham, lean, 3 oz.	250	70	High	High
Hamburger, lean, broiled 3 oz.	240	65	High	High
Veal, cutlet, 3 oz.	200	53	High	High
Poultry & Fish				
Broiled Chicken, 3 oz.	120	26	Moderate	High
Breaded Flounder, 3 oz.	140	39	Low	High
Herring (Pacific), 3 oz.	170	61	Moderate	High
Salmon, 3 oz.	120	38	Moderate	High
Tuna, 3 oz.	170	68	Moderate	High
Shrimp, 3 oz.	100	8	Low	High
Dairy				
Milk, whole, 8 oz.	150	49	High	High
Eggs, 1 medium	80	64	Moderate	High
Ice Cream, 4 oz.	140	48	High	High
Yogurt, plain, low-fat, 8 oz.	140	22	Moderate	Low

Content of Fat in Selected Food* (cont.'d)

	Calories Serving	% Calories From Fat	Saturated Fatty Acid Rating #	Cholesterol Rating #
Cheese				
Camembert, 1 oz.	90	73	High	High
Cheddar, 1 oz.	110	74	High	High
Cream, 1 oz.	100	90	High	High
Swiss, 1 oz.	110	65	High	High
American, 1 oz.	110	75	High	High
Cottage, creamed 1 cup	220	39	High	Moderate
Cottage, 1% fat, 1 cup	160	12	Moderate	Low
Nuts & Seeds				
Peanut Butter (Skippy Creamy) 2 Tbsp.	190	74	Moderate	None
Sunflower seeds ¼ cup	200	74	Moderate	None
Walnuts, 1 oz.	190	85	Low	None
Baked Goods & Desserts				
Angel Food Cake, 1 piece	100	1	Low	None
Apple Pie, ⅛ pie	300	38	Moderate	Low
Corn Muffin, commercial mix, 1	130	29	Moderate	Moderate
Devil's Food Cake, 1 piece	320	40	Moderate	Low
Fig Bar, 1	50	14	Moderate	None
Graham Crackers, 4	110	21	Moderate	None
Chocolate, 1 oz.	140	59	High	None
White Bread, 1 slice	75	11	Low	None
Donut, plain	170	41	Moderate	Moderate
Potato chips, 1 oz.	160	59	Moderate	None
Popcorn, 1 cup	40	41	Moderate**	None
Fruits & Vegetables				
Apple, 1 medium	80	9	Low	None
Avocado, ½	190	89	Moderate	None
Coconut, 1 piece, 2" 2" x ½"	160	92	High	None
Strawberries, 1 cup	60	11	Low	None
Watermelon, 1 wedge	110	8	Low	None
Potato, 1 medium baked	140	1	Low	None
Broccoli, 1 cup	50	11	Low	None
Beans, 4 oz. red kidney	110	7	Low	None
Corn, 4 oz.	90	8	Low	None

*Information in this chart is based on data from U.S. Department of Agriculture Handbook #456, *Nutritive Value of American Foods in Commun Units,* 1975 and "Cholesterol Content of Foods" by R. Feeley, P. Cirner and B. Watt, Journal of the American Dietetic Association, Vol. 61, pp. 134-9, 1972.

High, Moderate and Low ratings were assigned by comparing the levels of saturated fatty acids and cholesterol per 100 Calories of each food using guidelines based on current recommendations by many nutrition and health auth-

orities. Levels of 1.1 grams saturated fatty acids and 14 milligrams cholesterol for each 100 Calories in a food are rated Moderate. Significantly higher or lower levels are rated High and Low, respectively.

**Depends upon oil used in cooking.

Distribution of Fatty Acids in Common Fats and Oils

High in Polyunsaturates	High in Monounsaturates	High in Saturates	High in Saturates & Cholesterol
• corn oil	• peanut oil	• meat drippings	• lard
• safflower oil	• olive oil	• coconut oil	• lard-based shortenings
• sunflower oil	• margarines made with partially hydrogenated oil as the first ingredient	• palm oil	• butter and products made with dairy fats
• sesame seed oil		• hydrogenated shortenings	
• soybean oil			
• cottonseed oil	partially hydrogenated shortenings		• beef fat
• margarine with liquid oil listed as the first ingredient			
• mayonnaise			
• pourable salad dressings*			

*Read label for oil content and composition.

Note: These foods are classified by the fatty acids that are predominant in a given fat and oil.

By following this program, you will be introducing a powerfully *high concentration* of these *tart enzymes* that have the fat-splitting power to prevent accumulation within your system. It is the all-natural way to help protect against cholesterol overload.

"Meat-And-Potatoes" Eater Has Healthy Arteries

Max V. loved to eat heavy meat pies, thick stews, and favorite steak-and-potato meals. When his cardiologist told him that this lifelong love for heavy meats had raised his cholesterol levels to an unhealthy high and that he'd have to cut down, Max V. admitted it would be difficult. The food loves of a lifetime are not easily changed. He asked how he could enjoy his favorite foods and protect against atherosclerosis. His cardiologist outlined a simple program that would *control* fatty buildup and help keep his arteries reasonably clean.

Artery-Saving Program. Before each meal, stimulate your digestive enzymes by drinking a glass of grapefruit juice. This fruit juice, itself, is a powerhouse of strong enzymes. After each meal, have a citrus fruit platter or a seasonal melon half. *Benefits:* By strengthening enzyme power before the meat meal, and by boosting enzyme digestive power after the meat meal, there is more fat breakdown and elimination and less risk of a buildup. This program was followed by Max V. and his cardiologist later told him that he had reasonable cholesterol levels and healthy arteries…thanks to this enzyme fat-splitting program.

HOW TO USE ENZYMES TO PUT "GO POWER" INTO YOUR HEART

Raw food enzymes are able to exert a unique reaction within your body that will do more than search out and dissolve excessive fats. A powerful function of these enzymes is in the unparalleled power to stimulate the flow and maintenance of what are called *high density lipoproteins* or HDLs.

Important. This enzyme catalyst action causes an increase and then a sustaining level of these heart-saving substances. The amounts of HDLs that you have available for heart health may determine your actual lifeline. How to boost internal production of heart-saving HDLs? With the use of enzymes! If there is a deficiency, then the opposite occurs. There is an increase in levels of *low density lipoproteins* or LDLs.

HIGH-DENSITY LIPOPROTEINS VS. LOW-DENSITY LIPOPROTEINS

Yes, there is a vast difference. A heart- and life-saving difference. Let's take a closer look and then see how enzymes can help you create a needed balance.

What Are Lipoproteins? Substances within your body that are important for heart and artery health. Since fat and water do not mix, your body needs a method for carrying fats (including cholesterol) through the watery environment of your bloodstream. Proteins serve this purpose for cholesterol. There are two basic classes of cholesterol-protein aggregates. One is called *high density lipoproteins* or HDLs. The second is called *low density lipoproteins* or LDLs. Basically, they are produced by your intestine and liver and then released into your bloodstream. Here, through an enzyme-produced magnetic action, they become attached to cholesterol. Now what happens?

Note These Differences. To begin, understand that there are vital differences between the HDLs (the good substance) and the LDLs (the undesirable substance). The levels you have are determined by the enzymes you feed into your system. Note these internal responses.

HDLs. Energized by enzymes, HDLs act as cleansers wherein they help your body carry fatty substances through your bloodstream. Propelled by enzymes, these HDLs take excess cholesterol away from your blood vessel walls and other tissues and then aid in removing it from your body.

LDLs. If there is an enzyme deficiency, then the body appears to develop more LDLs which act much like a delivery truck, picking up cholesterol and then depositing it in your blood vessels and trillions of body cells. This may contribute to atherosclerosis, fatty deposits on the channels and pipes of your heart, and raise levels of cholesterol.

Enzymes: Hope for Desirable High-Density Lipoprotein Levels

When you eat a raw food, the released enzymes immediately start to break down ingested food. In particular, the raw food enzymes cause digestive enzymes (pepsin and hydrochloric acid) to break down and help dissolve accumulated fatty contents, along with other nutrients.

When this is done, the raw food enzymes stimulate the production of important blood proteins and this simultaneously will boost the increase of your desirable high density lipoprotein (HDL) levels. In so doing, this acts as a *natural protective factor* against cholesterol overload.

Helps Wash That Fat Right out of Your Body. The raw food enzymes then stimulate your HDLs to search out, take up, break down and prepare for elimination the fat that you may have eaten with your meal...and also the excess fat that may be threatening to clog your cells and choke your arteries.

Follow This Basic Program. *Before* you eat any meal, plan to boost your HDL levels by having a raw fruit or vegetable salad. *After* you finish your meal, insure continuance of raised HDL levels by having a raw seasonal melon. Again, enzymes will be introduced to work upon food that has just been ingested. This helps prevent and protect against excessive fats being stored. This basic program can do much to keep your arteries clean, melt away body fat and give the "breath of life" to your heart.

TWO FRUITS THAT ADD YEARS TO YOUR HEART

Two tasty fruits are able to send forth a shower of enzymes that will not only raise your needed HDL levels but *keep* them at a desirable level. These are —

1. *Papaya.* A luscious reddish-orange, this fruit is sweet with a musky flavor. Yet, it is a powerhouse of *papain,* a dynamic fat-splitting enzyme that is also able to exhilarate the release of important high density lipoproteins in your bloodstream. At the same time, the *papain* enzyme breaks up fat globules, and creates an artery-cleansing reaction that is like opening up channels of oxygen to your heart. Eaten regularly, this delicious fruit can actually add years to your heart.

2. *Pineapple.* Here is a sunshine-colored fruit that has a tangy and juicy good taste. The pineapple is the source of one of the most dynamic fat-splitting enzymes, *bromelin.* When you enjoy a slice or two of pineapple, you instantly release *bromelin* into your digestive tract. Here, this enzyme is able to cause a speedy metabolic reaction so that fats and proteins are broken up and distributed without excessive storage.

Improves Heart Health with Enzyme Fruits

Janet W. was told by her physician that she had a slight condition of cardiac arrhythmia (variation from the normal rhythm of the heartbeat). She had unfavorable cholesterol readings. Janet W. enjoyed her usual animal source foods. How could she have the best of all possible worlds by eating heartily and having a healthy heart? Her doctor told her to trim off all visible fats before cooking and before eating. But he also suggested that *during* each meal, she have a platter of both papaya and pineapple slices. Janet W. followed this program for three weeks. A new examination showed a healthy heart and, even more important, beneficial cholesterol readings. Thanks to the enzymes in these two fruits (she also ate fresh fruits and raw vegetables daily), she felt she had been given a new lease on life...while she continued enjoying her usual favorite foods.

YOUR ENZYME "HEART-SMART" FOOD PLAN[1]

Plan to cut down on fat and cholesterol. At the same time, increase intake of fresh raw fruits and vegetables as well as their juices. Follow this basic program:

1. To "butter" a baked potato or slice of toast, dip a clean pastry brush in vegetable oil and apply liberally. The taste of fat is there, but most of it is unsaturated.

2. Make hamburgers with lean chuck or lean ground round rather than fatty ground meats. Ask your butcher to grind only lean meats. Or, buy chunks of lean meat and grind it yourself. You might add a raw potato or a tablespoon of vegetable oil to the chopped meat to produce hamburgers that are juicy but low in saturated fat.

3. If a baking recipe calls for one cup of butter or solid shortening, substitute 2/3 cup of any vegetable oil.

4. If a recipe calls for sour cream, substitute plain yogurt. Yogurt liquefies easily, so add it carefully. Do not beat or stir too much and do not overheat.

5. If a recipe requires eggs, use two cholesterol-free egg whites instead of one whole egg. This won't work if the recipe requires many eggs, but egg whites are a suitable replacement if only one or two eggs are to be used.

6. You can easily enjoy low-fat meat or poultry soups. Make the soup one day ahead in the usual manner. When it cools, skim off all of the visible fat that floats to the surface.

7. If a baking recipe calls for whole milk, substitute low-fat or skim milk. You might also replace the milk with your favorite fruit juice for a bonus in enzymes.

8. Read the labels on imitation dairy products. Many contain a saturated oil such as coconut or palm oil. The labels may list

---

[1] *Parker Natural Health Bulletin,* West Nyack, New York 10994. Vol. 8, No. 17, August 14, 1978. Available by subscription.

these substances as "pure vegetable oil." If you're on a low-fat diet, it's best to avoid imitation dairy products that contain saturated fats. Switch to skim milk.

9. Brown onions, mushrooms and other foods in soy sauce instead of butter or oil. This method produces almost fat-free food that tastes as if it were cooked in fat.

10. Eliminate half the fat and calories in poultry by removing the skin after it has been cooked. Younger, smaller chickens and turkeys are leaner, so try to use them instead of older, fatter birds.

11. Experiment with different kinds of unsaturated vegetable oils. Sunflower seed oil, safflower seed oil, corn oil, soy oil, peanut oil and others are available at many groceries and health stores. They're cholesterol-free, yet offer you the satisfying and "safe" taste of fat.

12. Preserve the essential ingredients in oils by refrigerating them after you open the container. This will also help preserve the oil's taste and fragrance.

HOW OIL ENZYMES HELP LUBRICATE YOUR ARTERIES

The use of free-flowing oils will help send forth a rich supply of vital enzymes along with polyunsaturates that will help control cholesterol overload. Furthermore, oils offer these substances to help you boost your levels of beneficial high density lipoproteins. This will also help lubricate your arteries and guard against the risk of atherosclerosis.

The Enzyme-Rich Oils. These are made from fruits, vegetables or seeds. But *they must be cold-pressed.* This ensures their enzyme content. Remember that heat will destroy enzymes and if oils are prepared with a heating process, the valuable fat-fighters are destroyed. Therefore, *read labels carefully.* The oil should have been made with a cold-pressed process.

Does it matter? Yes. Basically, by means of a hydraulic press, the fruits, vegetables, seeds, nuts, grains, beans, etc., are pressed so that their oils are extracted along with the enzymes that are found in the raw plant food. A mechanical expeller press is used to squeeze the oils out of the plant. Pressure is exerted on the oil seeds through a continuously revolving shaft against a fixed surface. Because this "expeller press" or "cold-pressed" method is used *without heat,* it produces oils that are as close to the natural state as possible. Such oils have the best color, fragrance, flavor and nutritional benefits. Such cold-pressed oils are chock full of essential enzymes. Remember to read labels and use those that have been prepared in this high-enzyme-retaining manner.

THE "HIGH FAT" WAY TO CLEAN
ARTERIES AND A HEALTHY HEART

Free-flowing cold-pressed oils are a prime supply of enzymes along with polyunsaturated fatty acids. They work to "scrub" away the accumulated deposits along your miles and miles of arteries. They help keep open the channels of oxygen and nutrition that flow to your heart. With a simple *substitution,* you may continue to enjoy your natural liking for fat and be rewarded with a debris-free body.

Fat-Loving Farmer Has Healthy Heart

Agriculturist Andrew L. F. had a lifelong love for fat in his foods. Even after his cardiologist told him he had "too high" cholesterol levels and also excessive amounts of fat in his system, he could not deny his taste buds their favorite tastes. He had the urge to continue eating lots of fatty foods. His cardiologist devised an amazingly simple plan for Andrew L. F. All he need do was reduce or even eliminate hydrogenated (hard) fats from his meal plan. He would substitute with high-enzyme polyunsaturated (soft or liquid) fats in the form of cold-pressed oils. Andrew L. F. followed this program. He felt just as taste-satisfied as always. A bonus was that his cholesterol levels were brought to a doctor-approved level. He showed a healthier heart on the electrocardiogram test. With the use of these oils, he could indulge in fats with no fear of a cholesterol overload.

Your Simple Program: Just use oil wherever and whenever the occasion calls for the use of hard fats! You'll boost your levels of beneficial high density lipoproteins. You'll also introduce a treasure of fat-controlling enzymes from the cold-pressed oils.

Basic Guidelines for Healthier Arteries

1. Reduce the amount of total fat in your meal plan. Cut down on foods high in saturated fat such as beef, pork, luncheon meats, butter, cream, ice cream and fat-high cheeses.

2. Go easy on high cholesterol foods, especially egg yolk, liver and other organ meats.
3. Consume an adequate amount of calories to achieve and maintain your ideal weight.
4. Eat fish whenever possible as a meat substitute. It is an excellent low-fat source of the polyunsaturates you need to maintain healthy cholesterol levels.
5. If you must have sausage, give it a bath: stab with a fork and boil in water, allowing the fat to escape. Drain water and brown in a skillet.

RECIPE REVISION

Look closely at your recipes and their ingredients. You'll find that:

The amount of fat or oil can often be reduced by as much as ¼ to ⅓ in recipes for baked products. There are many recipes for the same item. Compare them and choose the one with the least fat. Fats can be entirely left out of recipes for soups, stews, casseroles, and the like.

Since fat adds taste to recipes by absorbing flavors, you may need to compensate by adding more seasonings.

Reducing fats in recipes may also require you to use more liquid for a proper batter consistency.

Substitutions for certain ingredients can skim fat (and calories) out of your recipes. Consider these suggestions:

TRY THIS:	INSTEAD OF:
DAIRY FOODS	
skim or lowfat milk (add nonfat dry milk powder to the liquid milk for quiches and custards to prevent a watery texture)	wholemilk
dry curd or low-fat cottage cheese	regular creamed cottage cheese
part skim ricotta cheese	whole milk ricotta cheese
evaporated skim milk	evaporated whole milk
evaporated skim milk or whole milk	cream

TRY THIS:	INSTEAD OF:
DAIRY FOODS	
whipped evaporated skim milk	whipped cream
low-fat yogurt, yogurt cheese, buttermilk or "mock sour cream"	sour cream or mayonnaise
Neufchatel cheese or "imitation" cream cheese	cream cheese
skimmed or partially skimmed milk cheeses (mozzarella, St. Ortho, Jarlsberg, etc.)	regular whole milk cheeses (cheddar, Swiss, etc.)
low-fat processed cheese slices	regular processed cheese slices
OTHER PROTEIN FOODS	
poultry, fish, veal	fatty beef, pork and lamb
ground round or lean ground beef	regular hamburger or ground chuck
the leaner cuts of meat	fatty, well-marbled cuts of meats
water-packed tuna	oil-packed tuna
MISCELLANEOUS	
imitation or low-calorie mayonnaise	regular mayonnaise or "salad dressing"
diet margarine (may need to reduce liquids in recipes)	regular margarine or butter

Other substitutions in your recipes can replace saturated fats in your family's diet with polyunsaturated fats which, you remember, tend to decrease blood cholesterol levels.

SUBSTITUTE THIS:	FOR THIS:
Recommended vegetable oils (listed in order of preference): safflower, sunflower seed, corn, soybean, sesame seed, cottonseed	Hydrogenated vegetable shortening
Margarine with a liquid recommended vegetable oil as the first ingredient listed	Butter, or a margarine with a hydrogenated or partially hydrogenated vegetable oil listed as the first ingredient

SUBSTITUTE THIS:	FOR THIS:
3 tablespoons dry cocoa plus 1 tablespoon polyunsaturated oil or margarine	1 ounce (or 1 square) baking chocolate
Polyunsaturated fat non-dairy creamer	Regular coffee cream, creamers made with coconut oil
⅞ cup recommended vegetable oil and ½ teaspoon salt substitute	1 cup butter

RAW FOODS ARE THE FOUNDATION OF ENZYME-CLEAN ARTERIES

Daily, boost your intake of a variety of seasonal raw fruits, vegetables, juices, seeds, nuts, whole grains. They are prime sources of enzymes and essential nutrients you need to help you keep your arteries clean, your heart healthy and your lifeline a very long one, indeed!

SUMMARY:

1. Fats can be "good" or "bad" for your health. Note the differences.
2. Two everyday fruits form the basis for your enzyme-boosting and fat-fighting program.
3. Max V. is able to indulge in his "meat-and-potato" passion and still have healthy arteries with his enzyme program.
4. Enzymes have the power to raise your high density lipoprotein (HDL) levels and give you immunity against cholesterol excesses.
5. Two enzyme fruits are able to add years to your heart through their natural fat-scrubbing action.
6. Janet W. improves her heart health with the use of several every-day enzyme foods.
7. Follow the tasty 12-step program for your Enzyme "heart-Smart" Food Plan.
8. Farmer Andrew L. F. follows a "high fat" way to clean arteries and a healthy heart.
9. It's easy to make recipe revisions as explained on the charts. You can feast on fats…and enzyme-wash your arteries and heart at the same time.

CHAPTER 9

How Enzymes Free You From

Headaches in Minutes

Headache! It may strike occasionally or frequently. It can cause minor irritation or more serious debilitation. It is one ailment for which chemical pain relievers promise help but offer minimal and temporary relief. It is estimated that over $300 million yearly is spent on pain relievers that mask the symptoms but do not help heal the cause. About 8 out of every 10 adults take patent medicines for headaches at least twice a month. It becomes habit forming. Headaches continue to return again and again, with pounding regularity. The reason here is that the *cause* (toxic overload) has been overlooked as the symptoms are merely "drugged" to give the brief illusion of relief.

WHAT IS A HEADACHE? Simply speaking, it is defined as "a pain in the head." It may last several minutes or hours. It may cover your whole head or one side of it. Perhaps the front or the back of your head. The pain may be steady or throbbing, barely noticeable or severe enough to cause confinement. Some folks are troubled by occasional headaches. Some have them almost daily!

Medical Term: The medical term for headache is *cephalalgia,* taken from two Greek words meaning "a condition of head pain." We are told that "headache is not a disease by itself but a symptom." This gives us a clue as to the *cause* of the problem. Namely, there is an accumulation of waste matter and toxic debris that causes abrasive irritation to the various neurotransmitters in your brain and along your entire nervous system. This causes the *symptom* of the head pain. To correct it, a form of enzyme cleansing can be used to help loosen up, dislodge, dissolve and then eliminate

121

the nerve irritants. This helps correct the *cause* of the headache and offers hope for relief, often in minutes.

ENZYME SOOTHING HOME REMEDIES FOR
SEVEN HEADACHE TYPES

Yes, there are different headache types which will respond to specific enzyme remedies for soothing relief. Most of them will work within minutes. They create this healing by means of a catalyst reaction wherein they will help dissolve the nerve-irritating waste products and then prepare them for speedy elimination. Once these irritants are removed from your body, your nerves become calm and you feel comfortable relief from the head pains. Check your headache type and the enzyme catalyst healing remedy for speedy relief.

1. Sudden Headache

Cause: It strikes "like a bolt out of the blue" with shattering impact. It usually happens after you have eaten a highly chemicalized food such as a synthetic canned, dehydrated or frozen product. The chemicals create a corrosive reaction on your cranial and physiological nerve network and this creates *abrasion.* The reaction is that of a sudden headache.

Enzyme Remedy: Drink a glass of fresh celery, cabbage, lettuce or any green vegetable juice. The rich concentration of enzymes will take up the potassium, magnesium and natural copper and use these minerals to *neutralize* the nerve-erosion reaction of food chemicals in your system. Just one or two glasses of cool green leafy vegetable juice will help wash out these irritants and give you quick relief.

2. Blurring Headache

Cause: Neurological distress is triggered off in the reaction of visual blurring, emotional upset and confusion, a decline in usual alertness. The basic senses such as sight, hearing, touch, taste may also be blurred. The cause may be traced to a "toxic overload" in which an excess of cooked foods and a minimum (or absence) of raw foods has caused an accumulation of irritant-causing debris.

Enzyme Remedy

This same situation occurred to plague traveling auditor Martin O'C. He had to subsist on cooked foods when on the road and this led to accumulation of toxic debris in his digestive system. Whenever Martin O'C. feels this type of headache, he just follows the remedy suggested by a health spa nutrition expert. *Simple:* Prepare a plate of freshly washed *raw* fruit. These should be *seasonal* since their enzyme powers are at a peak when they are tree or vine ripened. Eat as much as is comfortably possible of these raw fruits. Within 20 minutes, enzymes are active in digesting accumulated debris and preparing them for elimination. The headache symptoms will subside momentarily. The basic senses now become youthfully alert again, thanks to fruit enzyme cleansing.

3. Neck Stiffness Headache

Cause: Holding neck and shoulder muscles in a stiff and unrelieved position causes a form of cellular congestion in this region. The muscles that connect your upper shoulders to your neck become impinged by this unrelieved posture. Cells then become "choked" as waste matter is lodged in the crevices, unable to be transported to eliminative channels through a free circulation. This also tends to inhibit and even close off the flow of important nutrients to your brain cells. Consequently, this "choking" will cause head pains.

Enzyme Remedy: Free your choked up enzymes by using moist heat. In so doing, you will help release them so they can create a catalyst action; that is, propel essential brain cell-feeding nutrients to various portions of your cranium and thereby ease reactions. Just stand under a shower and let a thick stream of comfortably hot water play over the muscles of your neck and shoulders. *Note:* Just 15 minutes will create such soothing relief, the headache will be gone before you step out of the shower!

4. Recurring Headache

Cause: There is a nutritional imbalance in your eating program. At times, you may be eating an adequate amount of raw foods, but you may be omitting your *daily* intake of these enzyme sources. As a result, there is a recurring buildup of irritation-causing wastes which create a grating reaction on your cells and tissues. This is a seesaw reaction which makes you feel good for a few days, only to feel headache-plagued in another few days.

Enzyme Remedy: On *alternate* days, plan for at least one raw food meal. It may consist of your favorite seasonal vegetables seasoned with a bit of apple cider vinegar and oil for dressing. *Chew* the succulent vegetables thoroughly to release their waste-dissolving enzymes. In so doing, you will be supplying your digestive system with a storage supply of enzymes that will continue to wash out wastes even if you skip raw foods for a day or so. This helps control the "yo-yo" effect of headaches; that is, one day they're here, the next day they're gone. This *alternate* raw food program gives you adequate protection against "toxic overload."

5. Localized Headache

Cause: Toxicity tends to drift toward one side of the body, often the eye, ear, one side of the head, shoulder, etc. This may be traced to faulty posture in which excessive and unrelieved strain is placed upon that part of the body. Often, if you lie down after a meal, your digestive organs cannot perform enzyme catalyst action on eaten foods because of your semi-reclining or flat position. This causes a "backup" of foods and the reaction is for congestion on one side of your body. This is particularly noticeable if you lie down on your side after a meal, however mild. It leads to a localized headache.

Enzyme Remedy

A simple program was followed by Anna Z., who complained to her physiotherapist about those painful headaches that always plagued one side of her body. Anna Z. was advised to *take a walk for 30 to 60 minutes after each of her main meals.* That was all! Following that remedy, Anna Z. ended her localized headache problem in a matter of minutes. They never again returned.

The benefit here is that digestive enzymes have more catalyst power when subjected to comfortable and mild rhythmic movement after a typical meal. In a vertical position, food remains in the digestive tract where enzymes can create better digestion. But if you sit or lie down (especially on your side) there is the risk of ingested food washing upward from whence it came and enzymes are unable to create catalyst digestion. This will lead to recurring localized headaches.

6. Tension Headache

Cause: Primarily caused by contractions of the muscles of the scalp and neck. Generally, there is an overlap between muscle and blood vessel that will lead to that feeling of congestion. There are emotional factors involved that cause "up-tight" musculature that leads to headache.

Enzyme Remedy: The goal here is to relax the swollen vessels that have "locked in" enzymes so that they cannot efficiently cleanse these channels. A simple, but speedily effective enzyme catalyst remedy is to use chopped ice. Or an old fashioned ice bag. Just apply it on your forehead. You may also switch to the top of your head and then to the back of your neck. About 5 to 10 minutes for each side will help create a reflex relief of the arteries around your brain and thereby unlock the grating-irritating wastes and send enzymes to the area for swift dissolution and elimination. *Quick Relief:* Just 30 minutes with the use of an ice bag will create this enzyme catalyst cleansing action. You'll enjoy soothing relief within that time span.

7. Sinus Headache

Cause: Respiratory and bronchial congestion will lead to the accumulation of waste matter in the sinus passages as well as nasal orifices. When these toxic wastes build up, they block free passage of essential nutrient-carrying oxygen to your brain. The reaction is in the form of what is called a sinus headache.

Enzyme Remedy: Increase your intake of citrus fruits as well as their juices. Emphasize intake of orange, grapefruit and tangerine juices daily. Squeeze a little lemon juice into these beverages taken singly or in combination. The natural volatile oils of these citrus fruits and their juices will help penetrate the congestion of your nasal passages and permit a free transportation of needed oxygen. The high concentration of enzymes in the citrus juices will help dissolve the waste matter and actually wash them out of your body. Soon, you'll breathe better as your headache just goes away. *Quick Relief:* Grate fresh lemon, lime, orange and grapefruit peels. Put in a small jar with a tight cap. Frequently, *inhale* the fragrances. The strong, almost pungent scents from these fruit peels will penetrate the nasal congestion and offer speedy relief from sinus-caused headaches.

DIETER'S HEADACHE? REACH FOR A NATURAL SWEET[1]

The empty stomach felt by many dieters who have cut back sharply on quantity of food may bring on a headache, the result of

---

[1] *Parker Natural Health Bulletin,* West Nyack, New York 10994. Vol 9, No. 12. June 4, 1979. Available by subscription.

low blood sugar. Since blood sugar transmits oxygen to your brain, a sudden reduction may create an oxygen shortage that may then cause a headache. What to do?

Avoid Candy. Rich in refined sugar and calories, candy forces your pancreas to release more insulin to digest it; however, following this spurt, your blood sugar drops sharply and suddenly, worsening your headache problem.

Reach for a Fruit. It's a natural! Fresh fruit is a snack rich in fructose, a natural crystalline sugar that helps raise your blood sugar without insulin interference. Absorbed slowly, it causes a gradual raising of your blood sugar levels. This alerts enzymes to send a steady supply of oxygen to your brain. Your headache is eased. Your hunger subsides. You feel good as you lose weight... without the problem of headaches!

QUICK ENZYME RELIEF FOR
"LOW BLOOD SUGAR" HEADACHES

Dieting or not, you may be the victim of "morning headache." You may also feel recurring headaches for no apparent reason. If so, you may be subject to low blood sugar or hypoglycemia, as it is called.

Problem: Your brain (where the headache is felt) needs glucose (sugar in the form in which it usually appears in the bloodstream) in order to function. If you have a glucose deficiency in your blood, the resulting condition is similar to that of oxygen deprivation. To meet your brain's demand, more blood is called for. Therefore, the blood goes pounding through your head at an accelerated rate. Result: a throbbing headache.

Another Problem: If you're under stress, or in a nervous condition, your adrenal glands are speeded up and actually "devour" sugar at a speedy rate. This further denies your blood the needed sugar, and headache is the result.

Speedy Enzyme Relief: Enjoy a snack of fresh fruit. Try a handful of nuts or seeds. These are high concentrations of enzymes that will use Vitamin C as well as the rich supply of protein to help create more needed glucose to feed your demanding brain. *"Instant" Reaction.* Within five minutes after you thoroughly *chew* and swallow these simple foods, your blood sugar levels are raised, your headache is eased and erased! Enzymes work swiftly to create this catalyst-causing pain relief.

ENZYME HEADACHE TONIC

How To Prepare: To a glass of milk (use skim, if dieting) add one-half tablespoon brewer's yeast, one-half tablespoon lecithin granules, four ounces of fresh apple juice. Stir vigorously. Blenderize, if possible. Drink slowly. *Within ten minutes,* your headache should subside. More important, you'll discover a new feeling of youthful vitality and energy of body and mind.

Enzyme Catalyst Benefits: The calcium of the milk is combined with the all-important B-complex vitamins of the brewer's yeast and the highly invigorating choline of the lecithin. Now, enzymes in the apple juice will accelerate their absorption in your system. Within moments, enzymes create a catalyst action so that the nutrients of this tonic will soothe your nervous system, help calm down your frazzled neurotransmitters, or "relay stations" of your brain. In this combination, you begin to feel welcome headache relief within moments. Just one glass will do the trick!

Super-Benefit: The fruit enzymes seize hold of the choline from the lecithin and use it to synthesize the hormone epinephrine (adrenalin) and also influence your nerve impulses. This same enzyme-dispatched choline is used to nourish the segments and cells and tissues of your brain and create a comforting relaxation of tension. At the same time, there is a regeneration of your brain cells. This gives you a feeling of youthful vitality that is enjoyed even more because your headache is gone.

Whenever you experience a headache, prepare this tasty *Enzyme Headache Tonic* and actually enjoy your way to freedom from pain within minutes!

SAY GOODBYE TO MIGRAINE HEADACHES WITH ENZYMES

The word "migraine" comes from the Greek word *hemikrania,* meaning "half the skull." This appears logical since this type of headache is often confined to just one side of the head. A typical migraine consists of prodromal (early) symptoms. These include flashing lights or flickering vision to be followed by a starter of a throbbing or pulsating headache. The throbbing pain is usually in the same location but may change its area in the head. Often, it may radiate out of the eye. Nausea along with "blind spots," double vision and drooping eyelid may be noticed. Normally bright lights cause flinching. Normal sounds seem too loud. There is a torturous ache on the affected side of the head. The migraine headache may endure for as long as 48 hours...even longer, without relief.

How Enzymes Erase Brain Irritation

The primary cause of migraine headache may be traced to *cellular contamination.* That is, an accumulation of waste materials that irritate your 12 billion (yes, *billion!*) cells. These cells function as batteries, transformers and switchers. If these cells, along with the components of your brain (cerebrum, cerebellum, brain stem), become clogged with toxic wastes that are not cleansed, then there is a grating, corrosive and often inflammatory type of irritation. This leads to the migraine headache.

Enzymes are required to create a catalyst reaction whereby there is a natural and healthful cleansing process performed on your brain cells. The enzymes need to be dispatched to the site of difficulty (particular part of head that aches) and actually "scour" and "scrub" away the festering grit that is responsible for irritation-causing migraine.

Aims at Ache-Filled Side. By means of what is called "biotic energy," enzymes are energized to become magnetically drawn to the side of the brain that is causing the ache. Here, these enzymes are able to split the molecules that hold the waste materials in clumps and prepare them for elimination. It is this magnetized "biotic energy" that makes enzymes a catalyst force in helping to cleanse the particular portion of the brain that is in need of relief.

Enzyme Migraine Tonic...It Works in Minutes

Ends Lifetime Migraine Problem

Michael X. suffered from increasing migraine headaches as the years went by. Medications caused serious side effects. He was resigned to get used to it (which was painfully impossible) when his nutrition-minded physician recommended a simple *Enzyme Migraine Tonic.* Michael X. took it. In minutes, the migraine pain subsided. In eight minutes, it was gone. A few weeks later, his next migraine occurred. Again, he took this all-natural tonic and the pain ended. Now, he no longer had to "get used to it." Instead, he got rid of his migraine...with enzymes.

Easy Tonic To Make. To a glass of fresh vegetable juice add one teaspoon of lecithin granules, one tablespoon of wheat germ, one teaspoon of brewer's yeast. Stir vigorously or (better yet), blenderize. Drink slowly. The taste is every bit as good as the relief you will experience...in minutes.

How Enzyme Migraine Tonic Provides Relief

The fresh vegetable juice is a powerhouse of enzyme catalysts. These dynamic substances now take the vital *choline* from the lecithin, combine it with the B-complex vitamins from the wheat germ and brewer's yeast and create this migraine-relieving reaction:

Enzymes speed the essential choline and B-complex vitamins to nourish your starved brain cells. A magnetic action directs them to the side of the brain that is affected with congestion-caused pain. Enzymes use the choline to nourish and cleanse the neurotransmitters of your brain so that they may now send and receive messages without the pain-causing blockage. At the same time, enzymes will use choline plus the B-complex vitamins to promote a more healthful flow of blood through the arteries and protect against congestion which interferes with brain health. This congestion is the root cause of your migraine and other headaches. It is the power of enzyme-propelled choline that protects against this problem.

Tonic Is a Dynamic Healer. The tonic is able to accelerate the effectiveness of choline to invigorate the nerve signal transmitter in your brain. *Important:* You must include a raw vegetable juice when taking lecithin or wheat germ or brewer's yeast. These last-named foods are inactive in the absence of enzymes. The raw juice will provide the enzymes required to supercharge these foods throughout your bloodstream. The same enzymes then create a catalyst reaction whereby the choline and B-complex vitamins are actually "lifted out" and "deposited" on the side of the brain that is migraine-affected. Relief is just a matter of moments away.

Suggestion: At the first sign of an approaching migraine, prepare this simple tonic. Drink it slowly. You'll be building resistance to the threat of a headache. The cleansing-nourishing action will help build more than resistance...but *immunity* to recurring migraine and other headaches!

ALL-NATURAL HEADACHE ELIXIR

Two Ingredients Conquer Pain…in Ten Minutes

To a glass of fresh fruit juice, add one teaspoon of wheat germ. Stir vigorously or blenderize. Drink slowly. You will discover your headaches becoming milder and milder almost from the first swallow of this tasty *All-Natural Headache Elixir*. Within ten minutes, the pain will be a thing of the past.

Enzyme-Healing Powers. In many situations, a headache may be traced to the problem of *vasoplasms* (spasms of a blood vessel) that cause visual disturbances, or the problem of *vasodilation* (stretching of the blood vessels.) To correct these problems, enzymes are needed to bring about soothing relaxation and retraction of the blood vessels.

The *Elixir* is a highly concentrated source of fruit juice enzymes. These will take out the Vitamin B$_3$ or niacin from the wheat germ and then create an important healing process. Enzymes will use this vitamin to soothe the blood vessels, to relax the nerves, and then to help calm down the blood vessels so they are no longer dilated. Within moments, the headache will be eased and erased.

Suggestion: At the first sign of an approaching headache, however mild, use the *All-Natural Headache Elixir*. It works in minutes and the headache will be gone quickly.

HOW TO "WAKE UP" YOUR ENZYMES AND ENJOY FREEDOM FROM HEADACHES

You may be enjoying your daily quota of enzymes through juicy good fruits, succulent raw vegetables and their juices. But you need to keep the enzymes in these foods in a condition of alertness. You need to "wake up" your enzymes so they are actively able to cleanse your trillions of cells, keep you healthy and free from headaches. Here is a set of suggestions that are easy, enjoyable and, most important, effective in your program to say goodbye to headaches.

1. *Take Regular Walks.* Prolonged desk sitting leads to cellular congestion which, in turn, will bring on a headache. Even if you're home most of the time, you run the risk of developing a sluggish circulation that chokes enzyme powers. You need to take regular walks. Whenever possible (at least 60 minutes or more per

day), get out in the fresh air. Walk in a comfortable stride. You'll help alert your enzymes and protect yourself against headaches.

2. *Keep Your Neck Moving.* Upper body cellular congestion as well as vascular constriction create a blockade against enzymes. This means your brain cannot be oxygen-nourished. Headaches may then strike. Free your neck from congestion so enzymes can get through, with their supply of brain nutrients. *Suggestion:* When you sit, keep your neck moving. Close your eyes. Now start to rotate your head slowly. First move forward, then to your right, then to your back, then to your left. Look upward at the ceiling. Look downward at the floor. In gentle movements, you'll open up congested passageways and help allow enzymes passageway to nourish your brain and protect against headaches.

3. *Rag Doll Exercise.* Loosen up "tight spots" that have locked enzymes in a trap! Free them with this simple exercise. Stand with your feet apart. Your entire stance should be relaxed. Now bend forward from your waist. Let your arms dangle loosely. Make believe you are a rag doll. Just "bob" your arms, your trunk, your upper body. Turn yourself into a rag doll. Let tension melt away. In ten minutes (or less), you will restore flexibility to your body parts. Now the "trapped" enzymes are liberated. They flow freely via your bloodstream to all body parts to create cellular scrubbing, the foundation for pain relief.

4. *Water As A Pain Reliever.* Looking for a natural aspirin? Try water! Just reach for a comfortably hot shower instead of an aspirin. Aim for the nape or back of your neck and between your shoulder blades. About five minutes with a needle-spray shower will help melt tight spots, promote better circulation and relieve the cellular clogging that is responsible for headaches.

5. *Drink Lots Of Liquids.* Parched and inflamed blood vessels in your head are often responsible for headaches. Enzymes in foods must have a liquid environment on which to "swim" to the site in need of healing. Therefore, you need to drink lots of liquids. Water is good. Fresh fruit and vegetable juices are superior because they not only offer plant juices, but are prime sources of enzymes. In this liquid environment, enzymes are then propelled along to the head's blood vessels and create a calming, cooling and relaxing reaction. *Remember:* keep your body amply hydrated in order for enzymes to "swim" to the area in need of healing. Otherwise, thirsty enzymes are weak enzymes!

6. *Cool Foot Soak.* Congestive headaches can be relieved if you soak your feet in comfortably cool water at a temperature between 50°F. and 70°F. This helps cause a reflex relaxation of tight blood vessels around your brain. Channels are opened. Enzymes are catalyzed along your bloodstream and blood vessel routes. They dispatch cleansing agents for the debris-covered brain cells. Most important, enzymes will be carrying brain-feeding nutrients that help ease and then erase the headache. This can often bring a sigh of relief within five minutes!

Headaches, whether once in a while or once too often, can be agonizingly painful. You can enjoy freedom from headaches in minutes with cellular cleansing, brain nourishment and the use of enzymes to create this form of total healing.

HIGHLIGHTS:

1. Note the seven different types of headaches and their all-natural enzyme remedies that work within minutes.

2. Martin O'C. conquered blurring headaches by eating a plate of freshly washed raw seasonal fruit.

3. Anna Z. ended a problem of localized headache with a very simple all-natural, totally-free program. She took a walk after each main meal.

4. Dieting? Have headaches? Try the natural sweet remedy for speedy relief.

5. Troubled with "low blood sugar" headaches? Enjoy the fresh fruit or nuts and seeds program for enzyme healing.

6. Try the *Enzyme Headache Tonic* for pain freedom in minutes.

7. Michael X. ended his "lifetime" migraine problem with an easy-to-make *Enzyme Migraine Tonic*. He was freed forever from head pains.

8. Ordinary headaches are ended with the *All-Natural Headache Elixir*.

9. "Wake up" your enzymes and enjoy headache-freedom with the six all-natural remedies available to everyone. Some are *absolutely free!*

CHAPTER 10

The EZE-CCC (Enzyme CarboCal Control)
Way to Speedy Weight Loss

Excessive body fat is a major health problem. If you have too much of it, you increase your risks of developing high blood pressure, heart disease, diabetes, respiratory infections, gallbladder disease and possibly some forms of cancer. And too much body fat puts extra strain on your muscles and joints. Therefore, you are aware of the problem of overweight, yet you cannot keep off those extra and unwanted pounds. What can be the reason for this inability to lose weight?

Problem: Fat Cells Need Enzyme CarboCal Control

The basic problem here is the root cause of your overweight. The fat is in your cells! When you take in an excessive amount of glycogen (digested form of sugar and starch), it becomes stored in the form of fat in your millions of *adipose* or storage tissues. This fat continues to build up in your cells and the pounds continue to increase, along with bulging inches around your waist and thighs.

Fat Continues To Accumulate. As you continue eating foods, the adipose-storage (fat) cells continue to become engorged. These cells do, indeed, become storage tanks. The gray cytoplasm (the runny body within) and other working segments of the cell are now pushed out to the rim in what is identified as a "signet-ring" shape. The nucleus is in the signet section. The yellow fat becomes the core of the cell. As weight increases, each cubic millimeter of the adipose-storage (fat) cells tends to accumulate more and more storage capacity. This continues on until excessive weight is seen in the mirror and on the scales.

133

Basic Cause Of Overweight. The intake of excessive amounts of food may be the basic cause. But the absence of sufficient enzymes needed to burn up excess carbohydrates and calories is the real cause of your overweight. If you are able to correct this deficiency, you then set off a unique *cell-slimming* reaction wherein enzymes get to the core of your adipose-storage (fat) cells and create a combustion action to help metabolize and burn up the stored carbohydrates and calories. This is the key to helping correct the basic cause of your overweight.

HOW ENZYME FOODS MELT POUNDS FROM YOUR BODY, INCHES FROM YOUR SHAPE

When you eat an assortment of fresh raw vegetables, especially *before* and *after* a meal, you set off a chain reaction within your metabolic processes that will help get to the adipose-storage (fat) cells. Here, enzymes will uproot, dissolve and then wash out stored up weight-causing carbohydrates and calories. This action helps melt pounds from your body, shrink inches from your shape.

The EZE-CCC (Enzyme CarboCal Control) Method

Raw foods contain highly-charged molecules and enzymatic formations that are considered as "biomagnomobiles." They are dispatched through your digestive system to enter the adipose-storage (fat) cells. Here, these "biomagnomobiles" as these enzymes are called, actually go into the *mitochondria,* or the power centers of your cells. Once they penetrate to this core, the enzymes now actually seize hold of the stored carbohydrates and calories and convert them into energy that can then be burned up. In this manner, the adipose-storage (fat) cells are *slimmed down.* Once enzymes slim down the cells, then weight starts to drop and inches start to shrink. This helps create a "no fault" type of weight loss...often within a few days.

Therefore, weight loss can be created when enzymes enter the chambers of your trillions of overweight cells and cause a metabolic reaction to burn up stored up carbohydrates and calories.

How to Begin Your Personal EZE-CCC Weight Loss Plan

How can you tell if you are "overfat"? Simple. Consult a weight chart. Such a chart tells you how your weight compares with that of thousands of other people with the same height and build. The weights represent average or typical weights, not necessarily ideal or healthiest weights. But these charts are indicators of excessive fatness, except for well-muscled athletes. Overweight folks generally have excessive fat. (See the weight charts at the end of this chapter.)

Next, try either or *both* of these tests to determine if you are "overfat."

1. *The Pinch Test.* Grasp your skin between your lower rib and your waistline. If you are pinching more than an inch in thickness, you're probably "overfat"...even if the weight chart tells you your weight is typical.

2. *The Mirror or "Eyeball" Test.* Take a good look at yourself, unclothed, in front of a full-length mirror. If you see a lot of ripples, bulges and flabby tissue, you are "overfat."

First Step: Control Intake of Carbohydrates and Calories

You need to reduce consumption of excessive amounts of carbohydrates and calories. You need to stop overloading your fat cells. You do *not* deny yourself many of your favorite foods under this EZE-CCC speedy weight loss program. In fact, you can indulge in a portion of pie or a slice of your favorite chocolate layer cake on occasion. But you need to compensate by boosting your intake of enzyme foods to metabolize the carbo-cal intake from these sweets. To give enzymes a change, however, you should not take in an excessive amount of these weight-causing foods. Enjoy such "forbidden" foods in moderation and in balance with a stepped up enzyme program.

The charts in this chapter tell you which foods are high in carbohydrates and calories. Let your conscience (weight scale) be your guide. (See pages 143 thru 153.)

One Enzyme Meal a Day Helps the Pounds Go Away

As a convention show hostess, Betty Ann O'M. was tempted by the buffet and snack tables. All the colorfully delicious foods could not be resisted. Therefore, she was obviously "overfat" no matter how hard she tried to diet. A visiting nutritionist and weight control specialist told Betty Ann O'M. that she could help slim down on a simple program. Out of her three meals a day, *one* had to be devoted solely to fresh raw vegetables. In addition, she was able to cut down on other foods to help prevent overload. Results? Within one week, she lost *two inches* from her waistline.

By the end of the second week, she measured a *five-inch loss.* All this on a simple one raw vegetable meal a day. There was almost *no sacrifice* of her appetite. Yes, she was able to cut down on foods, but she still indulged in her favorite goodies...and lost weight, on this EZE-CCC program.

How It Works: Eaten foods rich in carbohydrates and calories become transformed into glycogen by your liver. This is then dispatched to your adipose-storage (fat) cells to become deposited as fuel. To help "burn" up this glycogen, enzymes are required to create a "biomagnomobile" action. They are transported to your cells, enter the *mitochondria* and provide a "spark" to "ignite" the fuel-burning process that causes weight loss.

Note: Without the presence of enzymes, the fat cells become fatter and fatter. Did you ever feel "heavy" after a particularly starchy or fatty meal? That's because your fat cells are swollen up. They are bulky. They are sluggish. But enzymes act as energizers by dissolving the fatty accumulations and helping to wash them out of your system. You need to have these enzymes *always available* to perform this fat-melting process. That is why *just one meal a day,* however small, consisting of raw vegetables prepared in any desired combination, will give you "enzyme insurance" to protect against adipose-storage (fat) cell overload!

HOW TO INCREASE YOUR ENZYME REDUCING POWERS

To put force behind your fat-dissolving enzymes, you need to follow some simple guidelines. They're easy. They're slimmingly effective.

Learn How to Eat. Do this slowly. Chew thoroughly. The act of proper chewing alerts salivary enzymes, also boosts the release of your hydrochloric and related digestive enzymes. Furthermore, food chewing will release enzymes from your raw foods and this increases the vigor of these fat-melting catalysts.

Small Is Satisfying. Reduce portion sizes. Use fewer or smaller amounts of carbo-calorie laden "extras" such as margarine, sour cream, cream cheese and salad dressings. You'll soon readjust your taste buds and find that small is tastefully satisfying.

Enjoy Cookies, Candies...in Moderation. Enjoy a few of these goodies. But if you find you can't stop, then just have these allotted few...and no more in the house!

Should You Eat? Yes, a regular meal is important. You'll be less likely to overeat later on if you're not a meal skipper. But note this — "time to eat" signals are not always caused by hunger. Often, just seeing treats like nuts or candy makes you want to eat. Before you eat, ask yourself: "Am I really hungry?"

Mind Over Platter. Don't eat because you're bored, frustrated or lonely. Find other ways to relieve these feelings.

ENZYME FOODS THAT KEEP YOU SLIM

You can use enzymes to help melt-dissolve fatty carbohydrates and calories in minutes after a meal. The secret here is to use these enzyme foods in two simple but amazingly effective ways: (1) Before a meal (2) After a meal. *Benefit:* Enzymes are catalysts; that is, they promote action or change without altering their own status. Therefore, you need these enzymes *lying in readiness* to dissolve fatty foods about to be eaten. Also, you need these same enzymes to *digest eaten foods* that might otherwise become stored in your tissues as fats.

How to Enzyme-Away Your Weight. Have an assortment of seasonal *raw* vegetables available for this simple two-step program. The catalyst action will penetrate the inner chambers of your cells, uproot the stored up carbo-cal reserves, condense and then dissolve them so that they can be easily washed out of your body. This is the basic power of enzyme foods. Now...here is a list of these everyday fat-fighting foods:

Apple (especially tart variety), banana, beet greens, cabbage, carrot, celery, cucumber, endive, escarole, garlic, lettuce, mint, mushroom, mustard greens, parsley, parsnip, pepper (green or red), radish, tomato.

Enzyme-Catalyst Action. If you will thoroughly chew and swallow either one or a combination of these fat-fighting foods, you will introduce a rich concentrated treasure of enzymes into your digestive system. Almost immediately, they will be prepared to catalyze incoming fat and prevent your adipose-storage cells from becoming overburdened. The catalyst action works in a flash when enzymes come in contact with ingested foods. You will thus be able to maintain a slim figure (and take off excess pounds) via this cell-scrubbing action.

Bonus Reducing Benefit: Eat one or a combo of these fat-fighting enzyme foods *after* a meal. Again, within moments, the enzymes are set into action to dissolve and prepare for metabolic combustion the excess carbohydrates, calories and fats that you have just eaten. In this manner, you will be able to have the EZE-CCC (Enzyme CarboCal Control) way to speedy weight loss and control.

Enzymes: All-Natural Appetite Suppressants

Steamfitter Nicholas DeL., complained that his expanding waistline was caused by his "worked up" appetite traced to hard physical labor. He took chemicalized appetite suppressants. They made him dizzy, to the point where his safety was jeopardized on his job. Without these drugs, his appetite caused so much eating, he had a 49 inch waistline! He asked a slimmer co-worker (who worked harder than he did) what the secret was for his appetite control.

In a word, "enzymes." The co-worker would carry a brown bag containing either chunks or slices or the whole raw vegetable (from group listed above.) Whenever he felt an unnatural or uncontrollable urge to eat, he reached for these enzyme vegetables. A few radishes, a few chunks of celery, and the appetite was calmed. That was the "secret" of his all-natural appetite suppressant.

Nicholas DeL. tried it. Within two days, his eating urges subsided to the point where he lost a few pounds. Within seven days he had a 44 inch waistline. By the end of seventeen days, he was a youthfully slim 36 inches. Nicholas DeL. would soon be a healthy 30 inches (down from 49 inches!), thanks to the enzyme power of the raw vegetables.

Controls Appetite, Fills You Up, Slims You Down

Carry an assortment of fresh vegetable wedges. Chew thoroughly and swallow whenever you have the urge to eat something that you know will add pounds. The enzymes in the vegetables will cause a healthful cellulose-swelling within your digestive tract. That is, the fibrous segments of the vegetables will be moisture-filled by enzymes. You'll then have a comfortable feeling of fullness. You'll have less of a desire to eat. In effect, the enzyme catalyst action will control your appetite, fill you up and (because of less eating) slim you down!

Enzymes are all-natural appetite suppressants!

BOOST ENZYME POWER WITH BEHAVIOR MODIFICATION

Enzymes react very sensitively to your emotional behavior. If you eat hurriedly, or are in a nervous state, then enzymes cannot flow properly or digest-destroy fats properly. If you *mistreat* yourself in any way, enzymes become upset and are unable to catalyze thoroughly.

Your emotional attitude can do much to boost (or weaken) enzyme power. Therefore, you need to control external stimuli, for your enzymes' sake. You do this by following a behavior modification plan. Adjust your attitudes.

Example: "Psyche" yourself up in the morning by saying aloud, "I will let enzymes make me thinner and thinner." Repeat this before each meal. Repeat it silently as you s-l-o-w-l-y chew the enzyme foods. Repeat it aloud when you face a high calorie dessert. Make yourself ruler of your appetite. Your enzymes will work more harmoniously with your thoughts.

Now, try some of these enzyme-pleasing behavior modification programs:

1. Eat on smaller plates. Half portions will look full if served on a salad plate. Visually, you think it's a lot. This creates happier enzymes and better appetite control. NOTE: Want a chocolate cake? Try a small piece on a very small plate. It will be as satisfying as a larger portion, and less fattening. Less for enzymes to catalyze, too.

2. Going shopping? Have a prepared list. Don't cheat! When you stick to your list, you can avoid buying too much. Not having it around the house means less temptation.

3. Reach for your pillow instead of some food. When the urge to eat strikes you, strike back! Punch a pillow. *Hard!* Let all your frustrations evaporate with this punching bag technique. Yell out! (Don't let family or neighbors hear you.) Release stored up tensions. In a few minutes, you'll feel relieved. Your enzymes are now calmer. You'll be able to avoid overeating and have stronger enzymes, too. (Anger can drain them!)

4. Eat at regular mealtimes. Always do it sitting down at a table, with your place set, even if you're alone.

5. Eat slowly. Pause between mouthfuls. It takes your stomach 20 minutes to signal your brain that you're no longer hungry. Give it a chance.

6. Want to treat yourself well? Instead of food, reward yourself with a new hairdo, new compact, new book, fresh flowers, a smaller-sized garment that will fit when you reach your desired weight goal.

7. Forget what Mother told you. NEVER clean your plate. Always leave something on it. Put it away in a plastic dish for next day use.

8. Cook only as much as you will need for the meal you are getting ready to eat. If it's not cooked, it won't be tempting to have seconds.

9. Go into your kitchen ONLY when you need to prepare food. Otherwise, keep out of it.

10. Feel hungry? Eat raw vegetables. Keep them up front in your refrigerator so they'll always be speedily available.

11. Feel like sliding off your diet? Stand before a mirror — clothed or unclothed. (Or, first one, then the other.) Aloud, say to yourself, "I'm not satisfied with what I see! I've got to take off more pounds and inches. I'm not going to eat!" It's a great self-booster.

12. NEVER eat while watching TV or reading. You lose track of food intake when engrossed in something else.

13. Whenever possible, eat with other people. Conversation tends to dull the appetite, sharpen the fat-fighting power of digestive enzymes. It also slows down fast eating.

14. Paste a "fat" photo of yourself on your refrigerator. You'll have to look at it before you open the door.

15. Begin a meal by eating your favorite food. Why? Once you've indulged in this temptation, you'll have less of an urge to fill up on other foods.

By building these programs into your daily lifestyle, you will then be promoting a "happier" environment for your enzymes. Their catalyst powers of fat-carbohydrate-calorie melting will be increased. They will help you slim down more swiftly and...more permanently.

The Enzyme Way to Conquer Your "Sweet Tooth"

Julie A.H. noticed her expanding waistline and spreading hips. Exercise made her tired and could not give her slimming satisfaction. Her problem? The gnawing "sweet tooth." When it struck, Julie A.H. was seized with the uncontrollable urge to eat some candy, a piece of cake, a chocolate doughnut. Trouble was, she couldn't stop with one and that was the cause of her girth! A public health dietician heard her problem, and suggested this enzyme way to conquer her "sweet tooth."

How to Do It. Want a sweet? Can't resist the urge? Try a sour pickle! Just chew on one well-washed sour pickle. Enjoy its tart taste. Swallow it. One or two pickles will end that urge for a sweet...at the "cost" of a couple of calories.

Julie A.H. tried it. Instantly, she was able to control the urge. In a few days, she had lost two inches from her waistline. In one week, the scales showed an appreciable drop. Her hips and thighs were about three inches slimmer, too. The enzyme way did work!

How Does It Work? The tart, almost sour taste of the pickle tends to take the "sharp edge" off your taste buds, particularly in your tongue. Enzymes will use the pickle juice to "turn off" your sweet urge and thereby ease the desire to eat. It's the easy (and natural) way to use enzymes for appetite control and conquering of your "sweet tooth."

ENZYME TONICS FOR SPEEDIER WEIGHT LOSS

A combination of different fruit and/or vegetable juices, together with grains, will help accelerate the catalyst action in your digestive system so that the fat-fighting powers of enzymes are boosted.

When to Take. Preferably between meals. Food that remains "locked" heavily in your system will now be better metabolized and assimilated with the enzymes in these tonics.

Grapefruit Juice. The high Vitamin C content is used by the tart enzymes to create cellular cleansing, loosening of and dissolution of stored up fats, carbohydrates and calories.

Sauerkraut Juice. (Salt-free variety is recommended; available in health stores and special dietetic outlets.) A half glass of this juice offers potent minerals that are used by enzymes to promote more effective cellular cleansing.

Celery-Asparagus Juice. The enzymes and minerals in celery work with the alkaloids in the asparagus to create a natural diuretic action. This consists of dislodging and discharging carbohydrate-calorie-fat accumulations in your adipose-storage (fat) cells. NOTE: Use fresh asparagus. If canned or cooked, the alkaloid value is diminished; the natural substances formed by combining this vegetable with celery become weakened or lost. Alkaloids are enzymatic principles found only in living plants, composed of carbon, hydrogen, nitrogen and oxygen. This combo creates the cell-slimming action that is desired.

Carrot-Cabbage Juice. A prime source of Vitamins A and C which are taken by enzymes for the rebuilding of your trillions of cells. The combination tends to create a comfortable feeling of fullness (natural appetite suppressant) and is also waist-slimming.

Cucumber Juice. You may flavor with a bit of lemon juice for potently tart enzymes. The high potassium count of this vegetable is used by enzymes to create a very strong carbo-cal-fat dissolution that promotes slimness from within. A natural diuretic, cucumber juice will help you lose weight speedily and even permanently.

Lettuce-Tomato Juice. The acid-based enzymes in tomato juice will utilize the high iron-magnesium elements of the lettuce, to help create improved formation of healthy red blood corpuscles. This creates an important "exchange" factor. That is, the engorged or fat-swollen cells are slimmed down and replaced with new and healthier cells. This process helps maintain important enzymatic cleansing and slimming of your entire body.

Tomato Juice. (Obtain the salt-free variety if purchased in a can or jar. Otherwise, blenderize or juice the tomato, yourself.) Once digested, the high citric and malic acid content will be used by enzymes to invigorate your metabolic process and help scrub away stored up carbohydrates, calories and fats. You may add a bit of lemon or lime juice for a piquant flavor, and a tart enzyme that boosts the power of these ingredients.

Do you weigh too much? Are you "spreading out" too much? Then take advantage of raw food enzymes that are able to get to the root of the problem. Your cells need to be slimmed down. This is done by enzyme carbo cal control. You can actually eat and drink your way to a speedy weight loss!

WHAT'S THE HEALTHIEST WEIGHT FOR YOU?

Best Weight (in indoor clothing)

MEN

Height	Age 20-29	Age 30-39	Age 40-49	Age 50-59	Age 60-69
5'3"	125 lbs	129 lbs	130 lbs	131 lbs	130 lbs
5'6"	135 lbs	140 lbs	142 lbs	143 lbs	142 lbs
5'9"	149 lbs	153 lbs	155 lbs	156 lbs	155 lbs
6'0"	161 lbs	166 lbs	167 lbs	168 lbs	167 lbs
6'3"	176 lbs	181 lbs	183 lbs	184 lbs	180 lbs

WOMEN

Height	Age 20-29	Age 30-39	Age 40-49	Age 50-59	Age 60-69
4'10"	97 lbs	102 lbs	106 lbs	109 lbs	111 lbs
5'1"	106 lbs	109 lbs	114 lbs	118 lbs	120 lbs
5'4"	114 lbs	118 lbs	122 lbs	127 lbs	129 lbs
5'7"	123 lbs	127 lbs	132 lbs	137 lbs	140 lbs
5'10"	134 lbs	138 lbs	142 lbs	146 lbs	147 lbs

Source: Pacific Mutual
Life Insurance Company

Milk, Cheese, Cream, Related Products

	IN GRAMS			FOOD ENERGY
	Weight	Carbo.	Protein	Calories
Milk, whole, 3.5% fat, 1 cup	244	12	9	160
Milk, nonfat (skim), 1 cup	245	12	9	90
Milk, partly skimmed, 2% nonfat milk solids added, 1 cup	246	15	10	145
Buttermilk, cultured, made from skim milk, 1 cup	245	12	9	90
Cheese, natural, Blue or Roquefort type, 1 oz.	28	1	6	105
Cheese, natural, Cheddar, 1 oz.	28	1	7	115
Cheese, natural, Cottage, creamed, 1 12-oz. pkg.	340	10	46	360
Cheese, natural, Cottage, uncreamed, 1 12-oz. pkg.	340	9	58	290

Milk, Cheese, Cream, Related Products (cont'd)

	IN GRAMS			FOOD ENERGY
	Weight	Carbo.	Protein	Calories
Cheese, natural, Cream, 1 8-oz. pkg.	227	5	18	850
Cheese, natural, Parmesan, 1 tablespoon	5	tr	2	25
Cheese, natural, Swiss, 1 oz.	28	1	8	105
Cheese, Pasteurized processed, American, 1 oz.	28	1	7	105
Cheese, Pasteurized processed, Swiss, 1 oz.	28	1	8	100
Cream, half-and-half, 1 tablespoon	15	1	1	20
Cream, light, 1 tablespoon	15	1	1	30
Cream, sour, 1 tablespoon	12	1	tr	25
Cream, whipping, heavy, unwhipped, 1 tablespoon	15	1	tr	55
Imitation cream products, powdered creamer, 1 teaspoon	2	1	tr	10
Imitation cream whipped topping, 1 tablespoon	4	tr	tr	10
Milk, chocolate flavored drink (2% milk), 1 cup	250	27	8	190
Milk, malted beverage, 1 cup	235	28	11	245
Milk, dessert, baked custard, 1 cup	265	29	14	305
Milk, dessert, ice cream, reg. (10% fat), 1 cup	133	28	6	255
Milk, dessert, ice cream, rich (16% fat), 1 cup	148	27	4	330
Milk, dessert, ice milk, hardened, 1 cup	131	29	6	200
Yogurt (made from whole milk), 1 cup	245	12	7	150

Fruits and Fruit Products

	IN GRAMS			FOOD ENERGY
	Weight	Carbo.	Protein	Calories
Apples, raw, 1 apple	150	18	tr	70
Apple juice, bottled or canned, 1 cup	248	30	tr	120
Applesauce, canned, sweetened, 1 cup	255	61	1	230
Applesauce, unsweetened or artificially sweetened, 1 cup	244	26	1	100
Apricots, canned in heavy syrup, 1 cup	259	57	2	220
Avocados, California, 1 whole	284	13	5	370

Fruits and Fruit Products (cont'd)

	IN GRAMS			FOOD ENERGY
	Weight	Carbo.	Protein	Calories
Avocados, Florida, 1 whole	454	27	4	390
Bananas, raw, medium, 1 whole	175	26	1	100
Blueberries, raw, 1 cup	140	21	1	85
Cantaloupes, ½ medium melon	385	14	1	60
Cherries, canned, red, sour, pitted, water pack, 1 cup	244	26	2	105
Cranberry sauce, sweetened, canned, 1 cup	277	104	tr	405
Dates, pitted, 1 cup	178	130	4	490
Fruit cocktail, canned, in heavy syrup, 1 cup	256	50	1	195
Grapefruit, raw, medium, white, ½ grapefruit	241	12	1	45
Grapefruit juice, canned, white, unsweetened, 1 cup	247	24	1	100
Grapes, raw, American type, 1 cup	153	15	1	65
Grape juice, canned or bottled, 1 cup	253	42	1	165
Lemonade concentrate, frozen, diluted, 1 cup	248	28	tr	110
Oranges, raw, 1 whole orange	180	16	1	65
Orange juice, frozen concentrate, diluted, 1 cup	249	29	2	120
Peaches, raw, medium, 1 whole peach	114	10	1	35
Peaches, canned, heavy syrup, 1 cup	257	52	1	200
Pears, raw, 1 whole	182	25	1	100
Pears, canned, heavy syrup, 1 cup	255	50	1	195
Pineapple, raw, diced, 1 cup	140	19	1	75
Pineapple, canned, heavy syrup, 2 slices & juice	122	24	tr	90
Pineapple juice, canned, 1 cup	249	34	1	135
Plums, raw, about 2 oz., 1 plum	60	7	tr	25
Plums, canned, heavy syrup, 1 cup	256	53	1	205
Prunes, dried, uncooked, 4 medium prunes	32	18	1	70
Prune juice, canned or bottled, 1 cup	256	49	1	200
Raisins, seedless, 1 cup tightly packed	165	128	4	480
Raspberries, raw, 1 cup	123	17	1	70
Rhubarb, cooked, sugar added, 1 cup	272	98	1	385
Strawberries, raw, capped, 1 cup	149	13	1	55
Strawberries, frozen, 1 10-oz. carton	284	79	1	310
Tangerines, 1 medium	116	10	1	40
Watermelon, 1 wedge 4 x 8 in.	925	27	2	115

Vegetables
and Vegetable Products

	IN GRAMS			FOOD ENERGY
	Weight	Carbo.	Protein	Calories
Asparagus, cooked, 4 spears	60	2	1	10
Beans, lima, cooked, 1 cup	170	34	13	190
Beans, snap, green and yellow or wax, cooked, 1 cup	125	7	2	30
Beets, cooked, drained & peeled, diced or sliced, 1 cup	170	12	2	55
Broccoli, cooked, drained, 1 medium stalk	180	8	6	45
Brussels sprouts, 7-8 per cup, cooked, 1 cup	155	10	7	55
Cabbage, raw, finely shredded or chopped, 1 cup	90	5	1	20
Cabbage, cooked, 1 cup	145	6	2	30
Carrots, raw, 1 whole (5½ by 1 in.)	50	5	1	20
Carrots, cooked, diced, 1 cup	145	10	1	45
Cauliflower, cooked, flowerbuds, 1 cup	120	5	3	25
Celery, raw, 1 large outer stalk	40	2	tr	5
Corn, sweet, cooked, 1 ear	140	16	3	70
Cucumber, raw, pared, 1 10-oz.	207	7	1	30
Lettuce, raw, Boston type, 4-in. diameter, 1 head	220	6	3	30
Lettuce, raw, Iceberg type, 4¾-in. diameter, 1 head	454	13	4	60
Mushrooms, canned, solids & liquid, 1 cup	244	6	5	40
Okra, cooked, 8 pods	85	5	2	25
Onions, cooked, 1 cup	210	14	3	60
Peas, green, cooked, 1 cup	160	19	9	115
Peas, canned, solids & liquid, 1 cup	249	31	9	165
Potatoes, med., baked & peeled, 1 potato	99	21	3	90
Potatoes, Med., boiled, peeled first, 1 potato	122	18	2	83
Potatoes, French-fried, frozen, heated, 10 pcs.	57	19	2	125
Potatoes, mashed, milk & butter added, 1 cup	195	24	4	185
Potato chips, 10 chips	20	10	1	115
Radishes, raw, 4 small	40	1	tr	5
Sauerkraut, canned, drained, 1 cup	150	7	2	30
Spinach, cooked, 1 cup	180	6	5	40
Squash, Summer, cooked, diced, 1 cup	210	7	2	30

Vegetables
and Vegetable Products (cont'd)

	IN GRAMS			FOOD ENERGY
	Weight	Carbo.	Protein	Calories
Squash, Winter, baked, mashed, 1 cup	205	32	4	130
Sweet potatoes, baked & peeled, 1 potato	110	36	2	155
Sweet potatoes, boiled & peeled, 1 potato	147	39	2	170
Sweet potatoes, canned, 1 cup	218	54	4	235
Tomatoes, raw, 3-in. diameter, 1 tomato	200	9	2	40
Tomatoes, canned, solids & liquid, 1 cup	241	10	2	50
Tomato catsup, 1 tablespoon	15	4	tr	15
Tomato juice, 1 glass (6 fl. oz.)	182	8	2	35
Turnips, cooked, diced, 1 cup	155	8	1	35

Meat and Poultry

	IN GRAMS			FOOD ENERGY
	Weight	Carbo.	Protein	Calories
Bacon, crisp, 2 slices	15	1	5	90
Beef cuts braised, simmered, pot-roasted, lean & fat, 3 oz.	85	0	23	245
Beef cuts braised, simmered, pot-roasted, lean only, 2.5 oz.	72	0	22	140
Beef: Hamburger, broiled, lean, 3 oz.	85	0	23	185
Hamburger, broiled, regular, 3 oz.	85	0	21	245
Roast, oven-cooked, such as rib, lean & fat, 3 oz.	85	0	17	375
Roast, oven-cooked, such as rib, lean only, 1.8 oz.	51	0	14	125
Roast, oven-cooked, such as round, lean & fat, 3 oz.	85	0	25	165
Roast, oven-cooked, such as round, lean only, 2.7 oz.	78	0	24	125
Steak, broiled, sirloin, lean & fat, 3 oz.	85	0	20	330
Steak, broiled, sirloin, lean only, 2 oz.	56	0	18	115
Steak, broiled, round, lean & fat, 3 oz.	85	0	24	220

Meat and Poultry (cont'd)

	IN GRAMS			FOOD ENERGY
	Weight	Carbo.	Protein	Calories
Steak, broiled, round, lean only, 2.4 oz.	68	0	21	130
Chicken, flesh only, broiled, 3 oz.	85	0	20	115
Chicken, breast, ½ with bone, fried	94	1	25	155
Chicken, drumstick, with bone, fried	59	tr	12	55
Lamb chop, thick, with bone, broiled, 4.8 oz.	137	0	25	400
Lamb, leg, roasted, lean & fat, 3 oz.	85	0	22	235
Lamb, shoulder, roasted, lean & fat, 3 oz.	85	0	18	285
Liver, beef, fried, 2 oz.	57	3	15	130
Pork, Ham, light cure, lean & fat, roasted, 3 oz.	85	0	18	245
Pork, Luncheon meat, boiled ham, sliced, 2 oz.	57	0	11	59
Pork, Luncheon meat, canned, spiced or unspiced, 2 oz.	57	1	8	165
Pork, fresh, cooked, 1 chop, thick, with bone, 3.5 oz.	98	0	16	260
Pork, fresh, roast, oven-cooked, lean & fat, 3 oz.	85	0	21	310
Pork, fresh, cuts, simmered, lean & fat, 3 oz.	85	0	20	320

Meat, Fish & Shellfish

	IN GRAMS			FOOD ENERGY
	Weight	Carbo.	Protein	Calories
Sausage, Bologna, 2 slices	26	tr	3	80
Sausage, Braunschweiger, 2 slices	20	tr	3	65
Sausage, Deviled ham, canned, 1 tablespoon	13	0	2	45
Sausage, Frankfurter, (8 per lb.) 1 frank	56	1	7	170
Sausage, Pork links, (16 per lb.) 2 links	26	tr	5	125
Sausage, Salami, dry type, 1 oz.	28	tr	7	130
Sausage, Vienna, canned (7 per 5 oz. can) 1 sausage	16	tr	2	40
Veal, medium fat, cooked, boneless, cutlet, 3 oz.	85	0	23	185
Veal, medium fat, cooked, boneless, roast, 3 oz.	85	0	23	230

Meat, Fish & Shellfish (cont'd)

	IN GRAMS			FOOD ENERGY
	Weight	Carbo.	Protein	Calories
Fish & Shellfish, Bluefish, baked with fat 3 oz.	85	0	22	135
Fish & Shellfish, Clams, raw, meat only, 3 oz.	85	2	11	65
Fish & Shellfish, Clams, canned, solids & liquid, 3 oz.	85	2	7	45
Fish & Shellfish, Crabmeat, canned, 3 oz.	85	1	15	85
Fish sticks, breaded, frozen, 10 sticks	227	15	38	400
Haddock, breaded, fried, 3 oz.	85	5	17	140
Ocean perch, breaded, fried, 3 oz.	85	6	16	195
Oysters, raw, meat only, 1 cup	240	8	20	160
Salmon, pink, canned, 3 oz.	85	0	17	120
Sardines, Atlantic, canned in oil, drained, 3 oz.	85	0	20	175
Shad, baked with table fat & bacon, 3 oz.	85	0	20	170
Shrimp, canned, drained, 3 oz.	85	1	21	100
Swordfish, broiled with butter or margarine, 3 oz.	85	0	24	150
Tuna, canned in oil, drained, 3 oz.	85	0	24	170
Eggs, whole, without shell, raw or cooked, 1 egg	50	tr	6	80
Eggs, scrambled with milk & fat, 1 egg	64	1	7	110

Grain Products

	IN GRAMS			FOOD ENERGY
	Weight	Carbo.	Protein	Calories
Bagel, 3-in. diam., 1 bagel	55	28	6	165
Biscuits, baking powder, 2-in. diam., 1 biscuit	28	13	2	105
Bran flakes (40%), 1 cup	35	28	4	105
Breads: Boston brown, 1 slice 3 x ¾ in.	48	22	3	100
Cracked wheat, 18 slices per loaf, 1 slice	25	13	2	65
Raisin, 18 slices per loaf, 1 slice	25	13	2	65
Rye, American, light, 18 slices per loaf, 1 slice	25	13	2	60
White, enriched, soft-crumb, 18 slices per loaf, 1 slice	25	13	2	70

Grain Products (cont'd)

	IN GRAMS			FOOD ENERGY
	Weight	Carbo.	Protein	Calories
White, same as above, toasted, 1 slice	22	13	2	70
White, firm-crumb, 20 slices per loaf, 1 slice	23	12	2	65
White, firm-crumb, 34 slices per loaf, 1 slice	27	14	2	75
Whole-wheat, soft-crumb, 16 slices per loaf, 1 slice	28	14	3	65
Whole-wheat, firm-crumb, 18 slices per loaf, 1 slice	25	12	3	60
Cakes—from mixes, Angelfood, 1 piece	53	32	3	135
Cupcakes, 1 small with chocolate icing	36	21	2	130
Devil's Food, 2-layer with icing, 1 piece	69	40	3	235
Gingerbread, 1 piece	63	32	2	175
White, 2-layer with chocolate icing, 1 piece	71	45	3	250
Cakes—from home recipes, Boston Cream, 1 piece	69	34	4	210
Fruitcake, dark, 1 small slice	15	9	1	55
Plain sheet cake with boiled white icing, 1 piece	114	71	4	23
Pound, 1 slice ½-in. thick	30	14	2	140
Sponge, 1 piece	66	36	5	195
Yellow, 2-layer with chocolate icing, 1 piece	75	45	3	275
Cookies—Brownies with nuts from mix, 1 brownie	20	13	1	85
Chocolate chip, home recipe, enriched flour, 1 cookie	10	6	1	50
Cookies—Chocolate chip, commercial, 1 cookie	10	7	1	50
Fig bars, commercial, 1 cookie	14	11	1	50
Sandwich, commercial, chocolate or vanilla, 1 cookie	10	7	1	50
Corn flakes, added nutrients, plain, 1 cup	25	21	2	100
Corn flakes, added nutrients, sugar-covered, 1 cup	40	36	2	155
Corn (hominy) grits, enriched, cooked, 1 cup	245	27	3	125
Corn muffins, made with mix, egg, milk, 1 muffin	40	20	3	130

Grain Products (cont'd)

	IN GRAMS			FOOD ENERGY
	Weight	Carbo.	Protein	Calories
Crackers, Graham, 2½-in. sq., 4 crackers	28	21	2	110
Crackers, Saltines, 4 crackers	11	8	1	50
Danish pastry (without fruit/nuts), 1 round pastry	65	30	5	275
Doughnuts, cake type, 1 doughnut	32	16	1	125
Farina, quick-cooking, enriched, cooked, 1 cup	245	22	3	105
Macaroni, enriched, cooked 'til tender, 1 cup	140	32	5	155
Macaroni and cheese, canned, 1 cup	240	26	9	230
Muffins, with enriched white flour, 1 muffin	40	17	3	120
Noodles (egg), cooked, 1 cup	160	37	7	200
Oats, puffed, added nutrients, 1 cup	25	19	3	100
Oatmeal or rolled oats, cooked, 1 cup	240	23	5	130
Pancakes, 4-in. diam., buckwheat mix & egg, milk, 1 cake	27	6	2	55
Pancakes, plain or buttermilk mix & egg, milk, 1 cake	27	9	2	60
Pie, 1/7 of 9-in. diam. pie, Apple 2-crust, 1 sector	135	51	3	350
Cherry, 2-crust, 1 sector	135	52	4	350
Custard, 1-crust, 1 sector	130	30	8	285
Lemon meringue, 1-crust, 1 sector	120	45	4	305
Mince, 2-crust, 1 sector	135	56	3	365
Pumpkin, 1-crust, 1 sector	130	32	5	275
Pizza (cheese) ⅛ of 14-in. diam. pie, 1 sector	75	27	7	185
Popcorn, popped with oil and salt, 1 cup	9	5	1	40
Popcorn, popped, sugar coated, 1 cup	35	30	2	135
Pretzels, Dutch, twisted, 1 pretzel	16	12	2	60
Pretzels, stick, small, 2¼-in., 10 sticks	3	2	tr	10
Rice, white, enriched, cooked, 1 cup	205	50	4	225
Rice, instant, ready-to-serve, 1 cup	165	40	4	180
Rice, puffed, added nutrients, 1 cup	15	13	1	60
Rolls, enriched, Cloverleaf, home recipe, 1 roll	35	20	3	120
Rolls, same as above, commercial, 1 roll	28	15	2	85
Rolls, frankfurter or hamburger, 1 roll	40	21	3	120
Rolls, hard, round or rectangular, 1 roll	50	30	5	155
Rye wafers, whole-grain, 2 wafers	13	10	2	45

Grain Products (cont'd)
Beans & Nuts

	IN GRAMS			FOOD ENERGY
	Weight	Carbo.	Protein	Calories
Spaghetti, enriched, cooked 'til tender, 1 cup	140	32	5	155
Spaghetti, tomato sauce, cheese, canned, 1 cup	250	38	6	190
Spaghetti, as above, home recipe, 1 cup	250	37	9	260
Spaghetti, meatballs, tomato sauce, home recipe, 1 cup	248	39	19	330
Spaghetti, as above, canned, 1 cup	250	28	12	260
Waffles, from mix & egg & milk, 1 waffle	75	27	7	205
Wheat, puffed, added nutrients, 1 cup	15	12	2	55
Wheat, shredded, plain, 1 biscuit	25	20	2	90
Wheat flakes, added nutrients, 1 cup	30	24	3	105
Beans, Navy (pea), cooked, drained, 1 cup	190	40	15	225
Beans, canned with pork & tomato sauce, 1 cup	255	49	16	310
Peanuts, roasted, salted, halves, 1 cup	144	27	37	840
Peanut butter, 1 tablespoon	16	3	4	95

Fats, Oils and Salad Dressings

	IN GRAMS			FOOD ENERGY
	Weight	Carbo.	Protein	Calories
Butter, reg. (4 sticks/lb.), 1 stick, ½ cup	113	1	1	810
Butter, reg. (4 sticks/lb.), 1 tablespoon	14	tr	tr	100
Butter, whipped (6 sticks/lb.), 1 stick, ½ cup	76	tr	1	540
Butter, whipped (6 sticks/lb.), 1 tablespoon	9	tr	tr	65
Fats, cooking, lard, 1 tablespoon	13	0	0	115
Fats, cooking, vegetable, 1 tablespoon	13	0	0	110
Margarine, reg. (4 sticks/lb.), 1 stick, ½ cup	113	1	1	815
Margarine, regular and soft, 1 tablespoon	14	tr	tr	100

Fats, Oils and Salad Dressings (cont'd)

	IN GRAMS			FOOD ENERGY
	Weight	Carbo.	Protein	Calories
Oils, salad or cooking, 1 tablespoon Corn, Cottonseed, Olive, Peanut, Safflower, Soybean	14	0	0	125
Salad dressings, Blue cheese, 1 tablespoon	15	1	1	75
Salad dressings, French, 1 tablespoon	16	3	tr	65
Salad dressings, Mayonnaise, 1 tablespoon	14	tr	tr	100

Source: U.S. Department of Agriculture.

IN REVIEW:

1. Weight gain is traced to obese adipose-storage (fat) cells. Reduce them with enzymes and you shed pounds and inches... in days.

2. Note the EZE-CCC method for slimming down your cells.

3. Betty Ann O'M. followed a "one enzyme meal a day" program and took off inches (and pounds) in days.

4. Check your desired weight on the chart. Also note the carbo-cal content of most foods. Be guided accordingly.

5. Nicholas DeL. used an all-natural appetite suppressant. It took off some 19 inches from his waist in less than three weeks.

6. Boost enzyme power with the listed behavior modification steps. They're fun. They're easy. They're slimming!

7. Julie A.H. was able to gain mastery over her "sweet tooth" with a simple food. Cost? About 15¢. Results? Priceless!

8. Drink your way to speedy weight loss with the variety of tasty enzyme tonics.

CHAPTER 11

How Enzymes Help You See Better,

Hear More Clearly

Sharpen your senses with the use of nutrition-carrying enzymes. Improve your powers of vision, increase your ability to hear better, when the sense-sharpening components of your eyes and ears are better nourished. This occurs when your body is fortified with an adequate amount of sense-sharpening enzymes.

Think of it this way: Enzymes are the "labor force" that builds and rebuilds your sense organs just as construction workers are the labor force that builds your house. You may have all the necessary building materials, but to create a strong house, you need workers that represent the vital life element.

So it is with your body. You may have all the important nutrients, but you still need the enzymes — the life element — to keep your body alive and well.

Enzymes are important in helping you see better, hear more clearly. Let's see how you can use everyday foods and enzymes to put more youthful sharpness into your eyesight, more audio power into your hearing.

ENZYMES FOR BETTER VISION

The sight-making portion of your eye, your *lens*, depends upon a regular supply of Vitamin A and C, along with other nutrients in order to create better sight. Enzymes will take these nutrients to nourish specific muscles surrounding your lens. Enzymes will use these nutrients to strengthen the muscle that surrounds your lens. In particular, the enzymes will invigorate this muscle so that it is able to contract or expand to change the shape of the lens. Then it will be able to "focus" more sharply. That is, enzymes create better "accommodation" so that your eyes can accommodate themselves to distant or near objects, large or small.

Enzymes Helpful For Older Folks. Enzymes are especially needed by older folks because this particular muscle surrounding the lens may lose its youthful elasticity. The ability to see things very close or very far away may become more restricted. This calls for strengthening of the sight-making components with the use of raw food enzymes. In older years, such a program is extremely beneficial. It may make the difference between wearing glasses or seeing without this aid.

Raw Juice Fast Boosts Eyesight Powers

In her middle 60's, Eleanor W.B. noticed her eyesight was dimming. Coming from a family where the wearing of eyeglasses was considered "heredity," she might have resigned herself to this "inevitable" circumstance. But her nutrition-minded ophthalmologist diagnosed her problem as an enzyme deficiency. He prescribed a one day a week raw juice fast. Eleanor W.B. tried it. During this one day, she would have large amounts of seasonally fresh raw fruit and vegetable juices. No solid foods. No other foods or beverages. Results? Within five days, she no longer needed a magnifying glass to read telephone numbers. She could read classified ads in the newspaper held at arm's length. She could see street signs across the street with the sharpness she enjoyed when a young person. Eleanor W.B. saved her dimming eyesight. She did not have to wear glasses. She could see more clearly than when she was a young person, thanks to the enzyme nourishment of her sight-making components.

How to Use Raw Juice Enzymes for Sight Improvement

Raw Foods To Use: Those that are high concentrates of important Vitamins A and C. Try carrots, squash, tomatoes, all green and yellow vegetables, apricots. Also enjoy cantaloupe, peaches, watermelon, oranges, grapefruits, tangerines, apples.

Juice Daily for Enzyme Power. Prepare a glass or two of fresh vegetable juice and drink during the early part of the day. Prepare a glass or two of fresh fruit juice and drink during the latter part of the day. You may want to comfortably consume up to six or eight glasses throughout the day.

One Day Per Week Offers Super-Enzyme Power. Plan to go on a raw juice fast at least one day per week. No solid foods. The benefit here is that without the interference of other foods, your digestive system is able to take hold of your juice enzymes, metabolize the nutrients almost completely within a few moments, and speed them to the components of your sight-making organism. *Without competition* or *dilution* of enzymes because of the presence of solid foods, there is a more *rapid* and *total* assimilation of the valuable sight-boosting nutrients. Enzymes are able to work more effectively in building and rebuilding healthier eyesight on this one day a week program.

LESS SUGAR + LESS STRAIN = STRONGER EYESIGHT[1]

Problem: Sugar acts as a destructive force upon nutrients, especially enzymes. To help nourish your organs of sight, it is important to cut down and even eliminate sugar in all its forms.

Solution: "You can improve your eyesight, especially your nearsightedness, by cutting down your intake of sugar and reducing prolonged strain," said optometrist Ben C. Lane, in a talk before the 1979 meeting of the Optical Society of America, an organization of physicists and scientists doing research in optics. "In addition," says Dr. Lane, "a mineral imbalance of chromium and calcium in the body can also create conditions that lead to possible nearsightedness."

Sugar Is Visual Problem. Laboratory tests were performed on 100 patients to check diet, eyestrain and hair samples. Based on his findings, Dr. Lane notes that the intake of sugar along with other refined carbohydrates may cause a depletion of the body's store of chromium.

"Sugar also forces the body to use its supplies of B-complex vitamins that are needed to regulate fluid pressure in the eye." Furthermore, "Folks with worsening nearsightedness consumed relatively more sugar and other refined carbohydrates than those with unchanging eyesight."

---

[1]*Parker Natural Health Bulletin,* Vol 10, No 3. February 11, 1980. Parker Publishing Co., West Nyack, New York 10994. Available by subscription.

Strain, Too, Causes Visual Weakness. Repeated, long eye-strain from doing close-up visual work, theorizes Dr. Lane, "may lead to an increase in eye fluid pressure which results in near-sightedness. If the close-up work is done day after day for more than a couple of weeks, fluid pressure increases as the eye starts to react," and this causes visual weakness; specifically the problem of near-sightedness.

Mineral Deficiency Is Probable Cause. Hair samples showed that patients with worsening nearsightedness have a "washing out" of calcium from the body. "One effect of eating too much "over-cooked protein' is to lose important B-complex vitamins which, in turn, leads to a calcium loss."

Relieve Fluid Pressure, Ease Nearsightedness

What is nearsightedness? Also called *myopia,* it involves a change in the outward or convex curvature of the lens of the eye. Normal eye muscles change the lens curvature constantly as you focus on objects at different distances. To focus on a nearby object, eye muscles elongate the eyeball to increase this curvature to enable light from nearby objects to focus on the retina.

Problem: In nearsightedness, this increased convexity becomes permanent, the result of an increase in the fluid pressure in the eye.

Natural Solution: Dr. Lane believes that if this fluid pressure could be regulated through elimination of sugar and easing of visual and muscular strain, the nearsightedness might be controlled, halted and, possibly corrected.

Basic Guidelines: Give up sugar in all forms. Boost your intake of calcium and chromium foods such as dairy products, salmon, deep green leafy vegetables, whole grains and brewer's yeast.

Important Enzyme Program. Be sure to include a *raw fruit* or *raw vegetable* when you eat any prepared or cooked foods. The enzymes will then be able to use the vitamins and the important calcium and chromium for nourishment of your sight-making components.

The Tonic That Gave New Sight to "Old" Eyes

Mark LeG. became troubled with increasingly poor eyesight. He was only in his 50's, yet he felt he had "old" eyes. His glasses required constant changing. Because he had a family history of glaucoma(serious condition in which fluid pressure within the eyeball is excessively increased), his physician said he was prone to that condition. He was then examined by an orthomolecular (nutrition-oriented) physician who noted that he had a serious nutritional deficiency, notably Vitamin A. But more serious was his low enzyme content. The two doctors suggested Mark LeG. follow a simple tonic to be taken daily. It would boost intake of Vitamin A with the use of enzymes. Mark LeG. followed this program. Within nine days, his vision sharpened. The fluid pressure returned to normal. He was now diagnosed as being reasonably safe from the threat of glaucoma. He drinks this tasty tonic daily as "eye insurance." He says it gave new sight to his "old" eyes.

Sight-Saving Tonic. In a 10 ounce glass, put 8 ounces of carrot juice. Add one tablespoon of cod liver oil. Add 2 ounces of green lettuce juice (especially freshly washed spinach). Stir vigorously. Drink slowly.

Enzyme Benefit: The enzymes in the carrot juice will take the high concentrations of carotene from that plant, transform it into Vitamin A. The enzymes then take the Vitamin A from cod liver oil and the lettuce (1/2 cup of spinach gives you over 7000 units of this vitamin) and use this nutrient to help nourish your sightsaving components. Just one glass per day will correct the deficiency and help your eyesight so that you should be able to see better.

Maintains Eye Fluid Balance. Enzymes will use the vitamin to siphon off excess fluid in the eye that may otherwise cause increased pressure within the eyeball and threaten blindness. But enzymes use Vitamin A to create a fluid balance so that there is a natural protective effect against this increase.

ENZYMES FOR BETTER HEARING

You are able to hear through a biological electrical process. That is, electric current is transported from the outside world to your nerves of hearing. Here, with the help of available enzymes, your nerves become activated and transport the sound throughout your audio mechanism and you hear distinguishable messages.

How Enzymes Help You Hear. Basically, your ear consists of three separate parts — the visible outer ear, the middle ear and the inner ear. The outer ear is a trumpet-shaped organ with a funnel that reaches into your ear drum. This is a thin, tightly stretched membrane. It vibrates when sound waves strike it (like the top of a drum.) The middle ear contains three small bones shaped like a hammer, an anvil and a stirrup. When sound waves vibrate the ear drum, this vibration is transferred via these tiny bones into the third part of your ear, your inner ear. In this portion, microscopically tiny hairs wave back and forth and in this manner conduct the sound to the nerve of hearing which transmits it electrically to your brain.

These functions take place in a flash. From the split second a sound is made, it is transferred through these three parts of your ear and sent to your brain.

The presence of sufficient amounts of enzymes determines the effectiveness of the sound you will hear.

Accepts Sound, Transmits It, Creates Understanding. Enzymes in your audio components alert your three segments to accept sound. Instantly, enzymes cause transmission and understanding as the sound becomes translated into a message in your brain. Were it *not* for enzymes, sounds would not be heard. They would be the same "lifeless" substances as foods you eat that are not digested. Therefore, you require an abundant amount of enzymes available not only in your body but in your daily food program so that they will be able to help you hear...and hear well!

The Vegetable Program That Restored "Lost" Hearing

In her middle years, Charlotte O'K. was faced with a progressively worsening hearing problem. At times, she had to cup a hand behind one ear to hear someone's words. Her family became upset over having to raise their voices to talk to her. Charlotte O'K. began to retreat into a world of what she considered to be frightening silence. Never to hear favorite music, never to talk to others, never to enjoy the sounds which made life worth living. She was terrified. Medications were of little use. She felt her condition worsening. She sought help from a holistic otolaryngologist, that is, a hearing specialist who treated the whole body rather than just the organ of hearing. He found that Charlotte O'K. was deficient in essential Vitamin A. He prescribed a simple program. Three times a day, she was to feature raw vegetables that would offer Vitamin A. She would drink their juices. She could steam other vegetables to make them more palatable. Furthermore, she could have a combination of fresh and raw vegetables daily. Results? Within six days, familiar sounds were restored. By the end of nine days, she had recovered her "lost" hearing. All this, thanks to the enzymatic power of raw vegetables.

How Vegetables' Enzymes Are Hearing Boosters. Fresh raw vegetables (and their juices) are prime sources of enzymes as well as the all important "hearing vitamin" or Vitamin A. A unique reaction occurs when you have an adequate amount of this combo.

Boosts Hearing. Enzymes will use Vitamin A to nourish your *cochlea,* a snail-shell-shaped spiral cavity, which is the seat of your organ of hearing. Now your *cochlea* is healthfully moisturized and also contains sensory nerves that must be energized by enzymes in order to transmit sound to your brain. Enzymes will use Vitamin A to nourish more than just your sensory nerves; they deposit needed moisture that makes it easier for messages to travel along enzyme-lubricated pathways to be transmitted to your brain. It is the enzyme-Vitamin A combo that creates this instantaneous reaction that helps you hear.

Deficiency Risk. If you are enzyme and/or Vitamin A deficient (both complement one another), then you run the risk of developing bone overgrowths of your *cochlea.* This growth may narrow or contract (squeeze) important nerves and blood passageways in your inner ear and impair hearing. A serious enzyme deficiency (even if you have adequate Vitamin A) will "dry" your cochlea and this leads to progressive hearing loss.

Raw Vegetable Juices Save Precious Hearing

Program: Twice daily, enjoy a glass of raw vegetable juices. You will be fortifying your hearing apparatus with valuable enzymes and speedily assimilated Vitamin A. While raw vegetables are succulent sources of these hearing nutrients (and should be included in your daily food program), for speedier results, *add* a glass or two of their juices.

Hearing Improvement Is Quick. Almost immediately, enzymes will take Vitamin A and use it to nourish the sensory receptor cells of your ear. Enzymes will work rapidly to use Vitamin A to help nourish and invigorate these same sensory receptor cells, making them more sensitive to sounds so that your hearing becomes more acute, and your senses much sharper. This helps give you overall hearing improvement. These raw vegetable juices may well be the natural way to restore "lost" hearing.

Raw Foods Improve Special Hearing Cells. When you thoroughly chew and digest raw vegetables, enzymes will take the Vitamin A to create a unique revitalization of one special part of your hearing organs. That is, enzymes will nourish your cilia cells, or tiny hair-like structures important to your hearing that need replenishment at all times. They depend upon Vitamin A for life just as you depend upon oxygen for survival! Enzymes will deposit Vitamin A right on these cilia cells which then become supercharged with sensitive vitality and are able to "hear" better...almost at once. So a raw vegetable program is vital for better and recovered hearing.

Where to Find Vitamin A Plus Enzymes

Raw food sources include carrots (in the form of carotene which is converted by your liver to usable Vitamin A), broccoli, sun-dried apricots, leafy green vegetables. Try any dark leafy green and deep yellow vegetable; also any deeply colored fruit. Use these as salads, as desserts and, of course, as juices. They are power-houses of the two nutrients that are the foundation for improved and restored hearing — enzymes and Vitamin A.

The hearing Health Shake That Restored Ear Strength

Stuart R.B. was under stress on his supervisory position. He never thought that unrelieved tension could be the cause of his fading sense of hearing. He became more ill at ease when employees, family and friends had to shout at him. He was, he thought, the victim of dreaded progressive hearing loss. He might have lost his hearing had not a company physician suggested a special program that would accomplish two simple benefits: (1) Ease his stress and nourish his tension-drained nerves (2) invigorate and nourish his hearing organs. Stuart R.B. tried the program. In three days, his "ear strength" was restored. Now he heard as well as a healthy youngster, thanks to the recommended *Hearing Health Shake.*

How To Prepare Simply. In a tall glass, add a mixed vegetable juice, preferably from carrots and yellow vegetables. Now add one teaspoon of lecithin granules (from health store or pharmacy), one teaspoon of brewer's yeast. Add 4 ounces of fresh citrus fruit juice. Add one teaspoon of ordinary wheat germ. Blenderize thoroughly. Sip, savor and enjoy. *How Often:* Twice a day. Plan for 10 ounces, each shake.

Eases Stress, Rejuvenates Cells, Improves Hearing

The easily prepared *Hearing Health Shake* is a powerhouse of enzymes. Now, within 15 minutes after you enjoy this beverage, these enzymes take the high concentration of B-complex vitamins and use them to help ease emotional stress. Enzymes use the B-complex vitamins to create improved glucose metabolism which is vital to your hearing. Enzymes will use the Vitamin B$_6$ from the lecithin and brewer's yeast as well as the wheat germ to create a stabilization of the supply of electrolytes found in the fluids of your inner ear. *Note:* Enzymes will use B$_6$ to help maintain an important balance of these electrolytes and thereby give you relief from stress. Also, you will find yourself hearing much better with this stabilization caused by enzymes.

Enzymes will use the Vitamin A from the yellow vegetables to nourish your cilia cells.

Vegetables + Lecithin = Better Hearing. This *Hearing Health Shake* is especially beneficial because of the combination. The Vitamin A *together* with the lecithin, taken at the same time, creates better digestion and assimilation of the essential nutrients. Enzymes can create this improved reaction with this combination.

The *Hearing Health Shake* is a good source of Vitamin C and E. Enzymes use both of these nutrients to cause an antioxidant action; that is, to see that your hearing cells receive an increased amount of oxygen. This is the key to better hearing...almost at once!

Relax Your Way to Better Hearing. This *Hearing Health Shake* is a prime source of B-complex vitamins that are used by enzymes to make you feel relaxed. This is the first step. Enzymes then use Vitamin A, B$_6$, C and E to help cause improved oxygenation of your hearing apparatus, a better glucose transport, a cleansing of debris in the small veins (capillaries) and thereby result in more relaxation...for more hearing!

You can actually drink your way to better health and better hearing! They do go hand in hand!

How to Put "Energy" into Your Hearing

Much hearing weakness may be traced to a form of glandular stress. Unrelieved tension will cause body levels of glucose to drop. Enzymes need to provide energy to your hormones so that you can receive more "energy" to hear better.

Basic Problem: Stress or tension will cause the hormone, adrenaline, to be released in an exorbitant amount. This hormone is needed to meet the challenges of stress in daily living. But unrelieved stress will cause an excess of adrenaline release. As a result, this substance will squeeze the vessels and arteries of the ear. Because circulation of the ear is the smallest of the entire body, when there is continuous stress, the ear is usually the first to weaken. Energy is siphoned to other body parts to meet situations of stress. The ear becomes fatigued. It starts to malfunction. It starts to lose hearing power.

Enzyme Solution: To provide more body energy so that there is a *balanced* release of adrenaline, boost your intake of enzyme-energy foods such as sun-dried figs, prunes, raisins, apricots; also enjoy fresh oranges and the juice of this golden enzyme-energy fruit. Eating one or two grapefruits a day will give you excellent fruit sugars that will control adrenaline release and provide you with sufficient amounts of glucose so that there will be no weakening of your hearing apparatus. By enjoying a daily supply of these enzyme-energy foods (and their juices), you'll be soothing your glandular system so that your feeling of stress will be eased. Your hearing will be more energized. You'll feel better all over.

Correct These Enzyme-Destroying Habits. Cut down or eliminate enzyme-destroying substances such as sugar, white bread, caffeine, salts, smoking, alcohol, hard fatty foods. They tend to destroy the environment in which enzymes thrive. They also cause a disruption of needed glucose levels, choke your circulation, reduce oxygen needed by your hearing apparatus. Such harmful foods and non-foods are antagonistic to enzymes. Eliminate them. You'll have stronger enzyme power and better sense health.

TWO ENZYME METHODS FOR RELIEF OF "RINGING IN THE EARS"

What Is Ringing in the Ears? It is called *tinnitus* (tiNEYE-tus), from the Latin for "tinkling." It is said to affect some 36 million people. It is "felt" by the unrelieved ringing, hissing, roaring (like a waterfall) sound in the ears. If neglected, it can cause not only annoyance but also progressive deafness.

Enzyme Deficiency Is Part of Cause. Basically, an enzyme deficiency is a nutritional deficiency. In this situation, this holds true for the hearing apparatus. To begin, within the auditory pathway there is a well-established reflex arc with fibers that conduct a steady discharge in both directions from one end to the other. These hair cells need nourishment. If deficient, then the "starved" hair cells or the neural pathways that transmit their output cannot function properly. The reflex arc is altered so that the perception of sound becomes distorted. The frequencies become deranged. This causes the maddening "ringing in the ears" known as tinnitus. Enzymes are required to create nourishment of these components so that sound transmission becomes *stabilized* and on an even level.

Enzyme Hearing Health Method #1

Wash Out Blood Fat Accumulations. High levels of blood fats cause red blood cells to clump together. This reduces the flow of oxygen to the inner ear where hearing is boosted and where the "ringing" causes disturbances. A raw vegetable plan will help melt down accumulated fatty deposits, cleanse the ear cells and help nourish the fibers so that the reflex arc is better able to transmit normal and not garbled or altered sounds. Plan to eat fewer "hard" fatty foods and more fresh fruits, vegetables and grains to wash away irritating fat.

Enzyme Hearing Health Method #2

Restore Electrolyte Balance. That "ringing" may be a "protest" by your metabolism against the electrolyte imbalance. (Electrolytes are naturally occurring ingredients in the liquids of your inner ear.) If you have either too many or too few such electrolytes, your hearing nerves seem to "shout" out in protest and this gives rise to the distressing sounds of "bells" or "waterfall" that keep you more nervous than ever.

To restore your electrolyte balance, you need to boost your intake of fresh raw vegetables. These are powerhouses of vitamins and minerals that are used by enzymes to help restore cell fluid balance.

Enzymes will use these minerals to establish electrical potential and permit more improved nerve conduction so that the electrolyte balance within your ear will be better balanced. In so doing, you should be able to (1) enjoy better hearing and (2) ease the painful problem of tinnitus.

With the help of enzymes and juices made from everyday foods, you can help sharpen your eyesight, improve your hearing and enjoy the best that life has to offer in the way of being able to see and hear your favorite things.

SUMMARY

1. Enzymes are important for your organs of sight and hearing.
2. Eleanor W.B. went on a simple raw juice fast that boosted her otherwise fading eyesight.
3. A doctor has found that less sugar and less strain and important mineral intake will strengthen eyesight. This program also boosts the supply of enzymes and improves basic eyesight.
4. Marc LeG. used an all-natural *Sight-Saving Tonic* and felt he was spared the sight-thief of glaucoma.
5. Your hearing depends upon an adequate intake of enzymes that are needed for nutrient metabolism.
6. Charlotte O'K. used a vegetable enzyme program that restored her "lost" hearing.
7. Stuart R.B. was faced with stress-relating hearing loss and the threat of deafness. He started taking a *Hearing Health Shake* and discovered its rich enzyme-nutrient content boosted his hearing in a matter of days.
8. Citrus enzymes are able to put "energy" into your otherwise "tired" hearing.
9. Relieve "ringing in the ears" or tinnitus with either (or both) suggested enzyme methods.

CHAPTER 12

Repair-Regenerate-Rejuvenate
Body Organs with
Enzyme-Collagen Programs

A little-known substance that is enzyme-manufactured within your body may hold the key to total health and self-rejuvenation. Having an adequate amount of this substance means that your vital glands and organs are able to be self-perpetuated, self-healing, self-rejuvenating. This substance depends upon enzymes in order to be manufactured. So we can see that enzymes together with this miracle substance can give you hope for "freedom from so-called old age."

COLLAGEN: THE "MISSING LINK" FOR SELF-REJUVENATION[1]

Known as *collagen,* this substance has the power to create internal as well as external healing; it promotes tissue and cell rejuvenation to give you a feeling of total health. Let's look more closely at this "missing link" and see how you can use enzymes to boost its production to give you the feeling of self-rejuvenation.

What Is Collagen? An enzyme-created protein-like substance made within your body. Collagen serves as a cement-like substance to hold together all the cells of your trillions of body tissues.

How Does It Boost Health? In addition to strengthening and rejuvenating your trillions of cells and tissues, collagen also nour-

---

[1]*Parker Natural Health Bulletin,* West Nyack, New York 10994. Vol. 9, No. 22, October 22, 1979. Available by subscription.

ishes and heals bones, body organs, blood vessels and glands. Enzyme-created collagen stimulates natural cell proliferation. It also creates and strengthens new cells and tissues as part of a never-ending cycle of internal healing.

What Are Deficiency Risks? Inadequate amounts of enzyme will cause an insufficient supply of collagen. This creates a "missing link" reaction in the healing chain of events, since this substance is needed to hold cells together. Without enzyme-created collagen, wounds do not heal; cells break down; tissues disintegrate; skin sags; bones, organs, the vascular system and glands start to break down. Furthermore, vitamins and protein cannot be fully metabolized so illness can occur.

How Can Collagen Be Produced Within The Body? With the intake of enzymes, Vitamin C, zinc and other elements, your body can produce this needed health substance.

How Will Enzymes Boost Collagen Formation? Found solely in raw foods such as fruits, vegetables, juices, seeds, nuts, whole grains, enzymes are needed to create a catalyst reaction. That is, enzymes will take raw materials from the body and bring about the birth of collagen so that your body cells and tissues are strengthened and rejuvenated. *Important?* Your body should have enzymes available at all time for daily collagen manufacture. A deficiency of enzymes may cause a deficiency of collagen and subsequent premature aging of your vital organs.

How Will Vitamin C Increase Collagen Supplies? Between your cells, Vitamin C is used by enzymes to build collagen that reinforces other intercellular material — in a way similar to the way that steel rods reinforce poured concrete. *Caution:* A Vitamin C deficiency causes a scurvylike condition in which tissues collapse, wounds do not heal properly, blood vessels hemorrhage and gums cannot hold teeth properly. *Remedy:* You need Vitamin C in order to protect against collagen disintegration and cellular destruction. You can obtain a powerful *combination* of both enzymes and Vitamin C in such foods as oranges, grapefruits, tangerines, lemons, limes, papaya, strawberries, cantaloupe, tomatoes, broccoli, green peppers, raw leafy greens, and freshly prepared juices from these foods.

How Is Zinc Helpful In Collagen Formation? During ordinary wear and tear, much collagen is broken down in your body.

Zinc is needed to invigorate the DNA and RNA substances which control cellular reproduction. Enzymes work with zinc to create these tissue-regenerating substances. A deficiency of zinc means that the DNA-RNA components become sluggish. Collagen manufacture slows down. The aging process follows. *Remedy:* Boost intake of lean meats, liver, whole grain breads and cereals, skim milk, fresh seafood. Your enzymes will take out the zinc (along with other valuable nutrients) from these foods and use them, together with Vitamin C, for the cell-tissue regeneration that will help protect against illness and aging.

How Essential To Basic Youthful Health Is Collagen? According to the British medical journal, *Lancet* (May 20, 1978), "Long regarded as inert, uninteresting and purely mechanical in function, collagen is attracting the close attention of physicians and biochemists. There is realization that a deficiency is involved in many diseases, from fatal heart and lung diseases to back pain and minor skin disorders. Presently, we recognize disorders primarily when biological malfunction occurs, but collagen may be involved in other diseases at a more subtle level."

Collagen Saves Life of "Helpless" Patient

The NHB refers to a male patient, diagnosed as having weak muscles. His skin was brimming with scars and hemorrhages. Mildly myopic, the diameter of his cornea was unusually small. His physician, Louis J. Elsas, M.D., at Emory University's School of Medicine in Atlanta, boosted his Vitamin C intake to increase his collagen formation. Otherwise, he would be considered "helpless" in terms of healing. Patient John A. was in a sad state.

Program: For 20 months, he was given 4000 units of Vitamin C daily.

Results? His healing and muscle strength improved. His corneal diameter grew. He was well on the way to recovery, thanks to the use of collagen. This was brought about by the use of enzymes which took Vitamin C and used it to create important collagen for basic healing.

Collagen: Key To Health: Collagen may be the "missing link" that influences rejuvenation through cellular proliferation. A basic program is to boost your intake of fresh raw foods, Vitamin C and zinc. This trio will help create the much-desired form of self-rejuvenation.

HOW ENZYME-COLLAGEN GIVES YOU HEALTHY GLANDS AND YOUTHFUL HORMONES

A healthy set of glands will create an internal "fountain of youth" through the flow of life-extending hormones. You need to keep your glands enzyme-collagen nourished so that they serve you properly. With the help of simple enzyme programs, your glands can function with youthful vitality. Let's see how you can create this enzyme-collagen glandular rejuvenation program.

How to Repair-Regenerate-Rejuvenate ("RRR") Your Glands

Pituitary Gland. A small gland that lies at the base of your brain. This gland controls almost all body functions through a network of releasing factors and hormones. They control your rate of growth, metabolic rate, electrolyte balance and (in women) ovulation and lactation. Because the pituitary gland affects your entire endocrine system, it is sometimes called the "master gland."

"RRR" Enzyme-Collagen Program. Combine fresh vegetables with a sprinkling of whole grains such as wheat germs or bran. The vegetable enzymes combine with the zinc and manganese of the grains, and create asynthesis and process of electron transfer. In so doing, the pituitary hormones are able to flow richly and abundantly. This helps you think better and enjoy overall improved health.

Adrenal Glands. Shaped like Brazil nuts, they sit astride each kidney. Each adrenal consists of two portions, the outer shell or *cortex,* and the core of *medulla.* When under stress, the adrenals manufacture hormones at an accelerated rate. The adrenals must have sufficient amounts of enzymes and collagen to synthesize such hormones as epinephrine, norepinephrine and corticosteroids. The enzymes rebuild and maintain your adrenals with collagen repair. Enzymes use collagen to repair these working glands. To maintain the pace to meet the challenges of daily responsibilities, your adrenals can function only if adequate supplies of enzymes and collagen are *always* available. Otherwise, the cells become "tired" and the entire emotional strength reserve begins to weaken. So-called "aging" may occur.

"RRR" Enzyme-Collagen Program. Boost intake of fresh citrus fruits and their juices as well as fresh deep-yellow and orange vegetables and whole grains. Enzymes will take the important Vitamin A for manufacture of adrenal-repairing collagen. Enzymes will use the Vitamin C to create cell specific components to keep your steroid hormones in youthful flow. Enzymes will also use collagen to heal and bind your adrenals so that they can release hormones to reduce inflammation and regulate your powers of immunity.

How Enzyme Foods Extended A Teacher's Lifespan

Although she was in her late fifties, Brenda V.W. looked much older. Her skin sagged. Her energy was weak. She had "raw nerves" and would lose her temper with her schoolchildren. She resisted the impulse to retire. She wanted to continue teaching. More important, she wanted to feel energetic and look younger, too. She discussed the matter with the school nurse. An enzyme-collagen program was outlined.

Daily Guideline: Before each meal, a raw fruit salad or platter to stimulate function of digestive enzymes, and to add additional enzymes. After each meal, a raw vegetable salad or platter. The salad enzymes would properly digest eaten food and release more needed collagen makers.

Skin Tightens, Energy Doubles, Nerves Calm

Within nine days Brenda V.W. responded with amazing vitality. She had a firm, youthful skin with energy and temperament to match. Now that she had her lifespan "extended," she continued teaching and was the delight of her pupils, her family and... herself.

How The "RRR" Program Worked. By making enzymes available, the teacher could create more collagen formation. This strengthens the adrenals. Enzymes protect against the formation of an excessive production of melanin, considered one cause of cell-tissue deterioration. With the use of enzyme-collagen healing of the distressed adrenals, there is a better protection against stresses of daily living. The entire body becomes enzyme-collagen rejuvenated and there is the feeling of total youth and health. Enzymes create this repair-regenerate-rejuvenate response in the body, often within minutes.

Thyroid Gland. A two-part endocrine gland shaped like a butterfly, it rests against the front of your windpipe. A healthy thyroid releases a *balanced* supply of the thyroxin hormone. This, in turn, stimulates the activity, or metabolism of your trillions of body cells. *Caution:* An enzyme-collagen deficiency will cause cellular deterioration of your thyroid. Erratic thyroxin release may occur. Your sick thyroid may release too much thyroxin to cause such speedy metabolism that you become unhealthfully thin, have rapid pulse and breathing, a risk of goiter as well as extreme nervousness. Too little thyroxin causes physical deformities, thickening skin, slowup of mental and physical reactions.

"RRR" Enzyme-Collagen Program. Enzymes need *iodine,* a valuable mineral, with which to mend and repair and maintain the health of your thyroid. Enzymes will use collagen and iodine to give you a healthy thyroid.

Natural Thryoid Tonic. To a glass of tomato juice, add one-quarter teaspoon of kelp (an iodine-rich sea salt available at most health stores), and juice from any green leafy vegetables. Stir vigorously or blenderize. Drink one glass daily. The powerful iodine supply will be activated by the presence of the enzymes and speedily repair and regenerate your thyroid so that a balanced amount of thyroxin is released. You'll soon feel the glow of warmth and youthful health.

Weight Is Normalized, Skin Firms Up, Energy Triples

Lois E. MacC. felt her weight either ballooning or dropping down to an abnormal low. Her skin became furrowed. Her energy was drained out. She thought she was ready for a rest home! An endocrinologist (gland specialist) diagnosed her problem as a malnourished thyroid gland. He suggested she try the *Natural Thyroid Tonic* every day. After five days, her weight became normal. Her skin smoothed out. Her body (and mind) energy not only returned but tripled in power! She felt "saved" from the gloom of a rest home! The tonic provided her with a powerhouse of enzymes that used the content of iodine to stabilize her thyroid. Furthermore, enzymes worked with collagen to repair the cells and tissues of this life-bestowing gland so that it became as good as new...and so did Lois E. MacC.

Pancreas Gland. A large, long organ, it is situated behind the lower part of your stomach. It consists of scattered islet cells (called the islands of Langerhans) which release insulin, the hormone needed to create proper metabolism of carbohydrates (sugars and starches). A weakening of the pancreas means there may be insufficient hormone manufacture and the problem of an insulin deficiency occurs. This may trigger off the problem of diabetes.

"RRR" Enzyme-Collagen Program. When you eat any protein food (assuming it is cooked) you should include a raw vegetable *at the same time.* By chewing and eating the protein plus vegetable *together,* you release important enzymes that are able to take the amino acids from the protein food and use it to rebuild the pancreas. The enzymes use amino acids for the manufacture of collagen, the substance that can make "new glands" out of "old glands."

In particular, the protein plus raw vegetable creates a unique regeneration action on your pancreas. The enzymes work to synthesize insulin in the beta-cells of the pancreas. The enzymes then cause the released insulin to attach itself to the cell membrane where it catalyzes the transport of glucose into the cell. It is this enzyme-collagen catalyst action that helps metabolize sugar in the body through the release and utilization of insulin.

You may eat a good protein food (meat, fish, eggs, dairy, peas, beans, nuts, seeds) but the nutrients remain "lifeless" without the availability of enzymes *at the same time.* That is why it is pancreas-regenerating to combine your protein plus enzymes when you plan your meal.

The Liver. Largest body gland, it weighs from 3 to 4 pounds. It is located in the right upper part of your stomach, just below your diaphragm. Its outer border may sometimes be felt just below your ribs. Your liver acts as a detoxifying organ. It filters out poisons, it manufactures about one pint a day of bile, which it then secretes. This process is vital to food digestion, especially of fats. A healthy liver will store carbohydrates in the form of glycogen for release into the bloodstream when required for other body parts, especially your muscles. Your liver is also involved in storing and distributing fat for daily use. A healthy liver can create a healthy body.

"RRR" Enzyme-Collagen Program. For these functions to occur, your liver needs to be cellularly strong. Furthermore, liver "energy" comes from enzyme catalyst reaction. It is essential to have a supply of enzymes in store and from daily food intake to see that the cells and tissues are constantly being repaired and regenerated. In so doing, enzymes keep your liver in a smooth functioning state of health. To boost and maintain *daily* enzyme availability, feed your liver high enzyme foods. These include very dark yellow and red vegetables, brewer's yeast, whole grains such as wheat germ and bran, cold-pressed oils to maintain proper lubrication, seasonal deep colored fruits such as peaches, apples, pears, berries, oranges, tangerines, grapefruits. Also, use the *juices* from these enzyme foods to further supply enzymes to build and rebuild the collagen of your liver and maintain organ strength...for body strength.

Daily Liver Tonic

As a machinist, Paul K.B. needed to keep alert both mentally and physically. But he noticed his reflexes were slow and he was getting tired, long before the five o'clock whistle sounded. He began to pale. His skin had a grayish pallor. He had "sour stomach." When he was examined by the company's internist, the diagnosis was that his liver needed nutritional mending. He was told to take a *Daily Liver Tonic.* Just one 8 ounce glass would help set off the "RRR" process so that his liver would be able to enjoy good health. So would he. After just five days of this simple *Daily Liver Tonic,* machinist Paul K.B. recovered. His reflexes were sharp. He felt energy enough to work overtime! His skin was bright. His stomach went from "sour" to healthfully "sweet." With a healthy liver, he was a healthy person again.

Daily Liver Tonic for Collagen Regenerated Liver

How To Prepare: In an 8 or 10 ounce glass, pour equal amounts of fresh orange juice and grapefruit juice. Now add one-half teaspoon brewer's yeast (from health store or pharmacy). Add one teaspoon desiccated liver (dehydrated liver with all connective fat and tissue removed). Add a half teaspoon of honey for flavoring, if desired. Thoroughly mix or blenderize. Drink one glass daily.

"RRR" Liver-Repair Benefits: The highly concentrated enzymes from the citrus juices immediately take the B-complex vitamins and high iron from both brewer's yeast and liver and prepare a form of collagen from the protein content. The enzymes actually propel the collagen onto the liver to repair-regenerate-rejuvenate this master organ...often in minutes. At the same time, the enzymes use collagen to strengthen the tubules and funnels of the liver. Now, enzymes help your liver in its detoxification proces (to guard against "sour stomach" which indicates a backup of waste products) and promote better cleansing of your body. Enzymes *alert-activate-amplify* the liver's ability to manufacture fat-digesting bile.

Special Benefit of Daily Liver Tonic. The enzymes will use Vitamin C from the citrus juices to cause an improved carbohydrate storage. This means that glycogen (stored carbohydrate) will be readily available for enzyme-release into the body to provide important energy. Folks who feel "always tired" may just need a booster in the form of released glycogen into the bloodstream. The high concentration of enzymes in this tonic will create this process...and you'll discover youthful vitality, in minutes, because of this unique reaction. Just one glass per day can help transform you from "always tired" to "forever energetic"!

The Kidneys. A pair of large glands in the small of your back. The function of the kidneys is to filter your bloodstream, help remove impurities and then release them as wastes in the form of urine.

"RRR" Enzyme-Collagen Program. You need to guard against problems of kidney stones. These form by crystallization of the dissolved salts from bits of wastes that find their way into the urine. You also need to have healthy kidneys so that wastes may be freely passed off. You can enjoy a healthy body when the billions of cells and tissues are collagen-regenerated. You have some 2,400,000 nephrons (channels through which wastes are expelled) which must remain in healthy condition at all times. Ordinary wear-and-tear of your body causes breakage and destruction of hundreds of thousands of these nephrons every day...almost every hour! You need enzyme-collagen available to guard against complete breakdown of your vital kidneys.

Boost your daily intake of carrots, cantaloupes, peaches, tomatoes, all green and yellow fruits and vegetables. Also include citrus fruits. You may eat regular favorite foods but be sure to include a raw vegetable and a raw fruit with the meal. That is, a raw vegetable at the start (to activate sluggish enzymes and introduce new ones) and a raw fruit at the end (to digest eaten food and create kidney-repairing collagen). This simple "RRR" Enzyme Collagen Program will help repair your nephrons so they can filter poisons out of your body more effectively.

Raw Food Program. To further improve your kidney's health, plan to go on a raw food program. Every five days, enjoy raw foods and their freshly squeezed juices. No other foods should be taken, lest they compete for enzyme attention and cause a slight weakening of collagen formation powers.

Benefits: When your metabolic system is treated exclusively to raw foods for one day, a powerful enzyme-collagen kidney-healing reaction occurs. Enzymes will use collagen to strengthen the *glomeruli* (tubules) segments of your kidneys. Enzymes will prompt the release of renin (a unique kidney enzyme) which then converts blood protein into a substance known as *angiotension*. This enzyme-prompted substance acts as a powerful vasoconstrictor of capillaries. When sufficiently open, they can promote improved oxygenated-nutritional circulation and thereby maintain healthy blood neutrality. This creates a condition of healthy kidneys...an internally clean body...and a feeling of self-rejuvenation. Just one raw food program, preferably every five days, can give you "forever young" kidneys and a "forever healthy" body.

Gallbladder. A pear-shaped pouch on the undersurface of your liver. A healthy enzyme-collagen nourished gallbladder will store and concentrate bile, which comes from the liver through a special duct and is delivered to the intestines through another bile duct. A problem here is that an unhealthy bladder may give rise to gallstones. These are *biliary calculi* and consist of substances (such as cholesterol) which have backed up and accumulated in the bladder. To protect against this risk, enzyme-collagen programs are required.

"RRR" Enzyme Collagen Program. Fresh raw fruit and vegetable juices are a powerhouse of enzymes and Vitamin C that tend to scrub away the wastes, liquefy the *biliary calculi* and protect against their becoming solidified or clumpy. Enzymes will use Vitamin C to manufacture collagen which then will nourish the gallbladder and strengthen the ducts through which bile flows. In so doing, there is a better "filtering" system and improved resistance against backup of accumulated stone-causing wastes.

"Bladder Washing Program"

Coming from a family in which gallstones were considered hereditary, Ida L. was determined not to fall victim to this so-called "inevitable" problem. Her four older sisters and her mother had all suffered from bladder distress and stones. So she was going to change the course of "heredity." She discussed the issue with a nutrition-oriented endocrinologist. He suggested she follow an easy *Bladder Washing Program.* Ida L. did just that. For three days, she would have nothing but fresh fruit and vegetable juices. That was all. Results? She was examined and pronounced healthy. Her bladder was clean. She had not the slightest trace of sediment. Ida L. has defied the theory that gallstones "run in the family." With this simple once-a-month, three day raw juice fast, she has enzyme-collagen rejuvenated her bladder, promoted internal cleansing and freed herself from the threat of gallstones!

Enzymes + Collagen = Young Organs

You consist of billions, even trillions, of cells. By keeping these cells enzyme-collagen repaired, you can continue enjoying youthful health. Your organs consist largely of cells. Keep these cells repaired and your organs are youthfully healthy. So will you be! Raw food enzymes are able to strengthen the outside cell membrane of your trillions of organ cells. This permits a free exchange of wastes for nourishing oxygen. Enzymes will boost the strength of your cellular substances DNA-RNA which hold the key to total health. Feed these life-giving substances daily amounts of enzyme-collagen with raw foods and their juices and you will be rewarded with youthful vitality...inside and outside.

MOST IMPORTANT POINTS:

1. You are as youthfully healthy as your body organs. Regenerate them with simple enzyme-collagen programs.

2. Collagen creation through enzymes can hold the key to your quest for total health of body and mind.

3. John A., considered "helpless," is saved with the use of collagen.

4. Repair your glands with the "RRR" programs outlined.

5. Brenda V.W. responded with amazing vitality on an enzyme-collagen program that revitalized her adrenals (and body, too).

6. Lois E. MacC. normalized weight, firmed up skin, tripled energy with a *Natural Thyroid Tonic.*

7. Machinist Paul K.B. enjoyed total rejuvenation on a *Daily Liver Tonic* that enzyme-energized his entire body.

8. Ida L. defied "heredity" which proclaimed that she would develop gallstones because they "ran in the family." On a high enzyme-collagen *Bladder Washing Program,* she cleansed this vital organ, boosted internal regeneration of her trillions of cells — and enjoyed a feeling of freedom from gallstones...despite "tradition."

CHAPTER 13

Free Yourself from Allergies with
New Enzyme-Catalyst Programs

The word allergy is derived from Greek *allos* plus *ergon*, meaning "something else at work." This is the result of cellular disintegration and decay. Your body, especially your respiratory tract, must *work harder* to take in oxygen and maintain your circulation and other metabolic processes. It is this *extra burden* that causes body assault and gradual breakdown. By building inner resistance to allergic substances through molecular fortification, you can help free yourself from these disturbances. In particular, an enzyme-catalyst program is important to help restore and regenerate your fragile and decaying cells and tissues. In so doing, you will build resistance to allergies.

Let's look at common and uncommon allergies and see how enzyme-catalyst programs can help you overcome their distress.

THE BIOFLAVONOID ENZYME WAY TO FREEDOM FROM ASTHMA

Basic Problem: Asthma is an ailment of the lungs in which you experience difficulty in breathing because of obstruction to the flow of air in the bronchial tubes. This is due to swelling of the lining membranes, contraction of the surrounding musculature and a plugging of the tubes by thick mucus. The rush of air through the narrowed tubes produces wheezing — the noisy, whistling sounds that are typical of asthma. You have to "work harder" to gasp for the breath of life.

Bioflavonoid Enzymes Strengthen Breathing Cells. With an adequate intake of bioflavonoid enzymes, you will be able to repair and regenerate your broken, fragile, disintegrating cells of your bronchial tubes. These bioflavonoid enzymes work swiftly to rebuild your *alveoli* (millions of tiny sacs with very thin skins) which make up much of your respiratory organs. The bioflavonoid enzymes then nourish your breathing *capillaries* (smallest of your blood vessels) with fragile skins that surround your alveoli. When both your alveoli and capillaries are enzyme restructured and strengthened, they can protect you against irritation from inhaled substances that trigger off an asthmatic attack.

Anti-Asthma Power of Bioflavonoid Enzymes

Basically, bioflavonoids belong to a group of nutrients believed to be super-powerful and much more effective than Vitamin C. These bioflavonoid enzymes are found in the *rinds* and *pulps* of citrus fruits, especially oranges, tangerines, grapefruits; you will find them in rose hips, most berries, black currants and, to a lesser degree, in sun-dried apricots, raisins, figs and dates. Grapes contain appreciable amounts of bioflavonoid enzymes.

Bioflavonoid Enzymes Regenerate Lungs. Almost immediately after eating these fruits (or their juices), the bioflavonoid enzymes go to work to repair the tiny blood vessels that make up your bronchial tubes and breathing components. The bioflavonoid enzymes repair the delicate alveoli and capillaries that make up these organs. This protects them from irritating substances that are inhaled.

Unique Anti-Asthma Power. You break out in an asthmatic attack because inhaled offenders are able to strike right through these delicate cells. This irritation causes the cells to produce antibodies which react with the allergen, to provoke the attack. Therefore, bioflavonoids are needed to strengthen these capillaries. In particular, the bioflavonoid enzymes work swiftly to strengthen the intercellular cement which is found in the walls of your alveoli and capillaries.

Once this happens, your cells and capillaries are strengthened so that there is a *natural shield* to *insulate* your bronchial tubes against the harsh and irritating effect of inhaled allergens.

Bioflavonoid enzymes are able to give you this *cellular resistance* to the irritation of the inhaled offenders. This gives you a natural and built-in fortification that is likened unto an anti-asthma power. It is available through daily intake of bioflavonoid enzymes.

Says "Goodbye" to Lifelong Asthma in Nine Days

Bus dispatcher Ned T.W. had a lifelong battle with asthma. At times, he would break out in such choking, wheezing attacks, he turned purple with strained efforts to inhale precious air. He took medication but it offered partial relief; also, he noted when the drug effects wore off, his asthmatic outbreaks increased in frequency. They worsened, too. He felt unhappy and downright miserable because of this never-ending asthmatic affliction.

Allergist's Program. His allergist suggested a new program that appeared to be effective in clinician trials. It consisted of mega-doses (very high amounts) of bioflavonoid enzymes. The allergist suggested that Ned T.W. eat a platter of bioflavonoid enzyme fruits every single day. At the same time, he would be able to enjoy the juices of these fruits throughout the day. *Special Program:* In the morning, breakfast should consist solely of these fruits. Beverages should be their juices.

Benefit: Prepare your respiratory tract for the day ahead. Strengthen the alveoli and capillaries with high concentrations of bioflavonoid enzymes to meet the challenges of the day to come.

Results: Within three days, Ned T.W. noticed a lessening of the severity and frequency of his asthmatic attacks. At the end of five days, he had just one very mild asthmatic outbreak per day. By the time he reached his ninth day, he was asthma-free. For the first time in his life, Ned T.W. was free from asthma! He had to thank his nutrition-allergist and, of course, the power of bioflavonoid enzymes!

Why Fresh Fruits Are Super-Powerful in Bioflavonoid Enzymes

A powerful high concentration of bioflavonoid enzymes will be found in the white underskin and segment part of the particular fruit. (Oranges, grapes, plums, black currants, grapefruits, sun-dried apricots, cherries, berries, tangerines, lemons, limes are all prime sources of bioflavonoid enzymes.) Therefore, you would do well to eat these fruits whole. If the fruit has to be peeled, do NOT remove the white layers under the skin or around each segment of fruit (such as the orange, tangerine, grapefruit.) These white layers (look at them) are prime sources of super-bioflavonoid enzymes.

Eat these white layers along with the fruit and you will be regenerating your trillions of alveoli and capillaries so that they can resist the irritation of inhaled offenders.

Simple Anti-Asthma Program: Each day, especially for breakfast, plan to include a variety of the aforedescribed bioflavonoid foods. Eat them daily. They will immediately work to repair your fragile capillaries and build natural resistance and freedom from asthma. You may also enjoy the juices of these fruits. While the bioflavonoid concentration may be less, the enzyme content is very high in juices so you should take them daily. This combination is your ticket to a world in which you can breathe more easily.

THE ENZYME CATALYST PROGRAM TO OVERCOME HAY FEVER

Problem: From early spring until the first frost in autumn, about one in fifteen persons suffers from what is called hay fever. Depending upon the section of the country, hay fever comes with amazing regularity. *Symptoms include:* spasms of sneezing, stuffed-up, runny and itchy nose, itchy and swollen eyes; there is itching and mucus in the throat. *Caution:* neglected hay fever becomes worse as the years go on and may then develop into asthma. That is why it should be overcome as early as possible.

The term *hay fever,* by the way, is a misnomer. It is *not* caused by hay. It is *not* associated with fever. The correct name should be *seasonal allergic rhinitis.* But it is commonly called hay fever so we refer to it as such.

Enzyme Fasting Program Clears Up Hay Fever

Basic Cause: Hay fever occurs when airborne pollen and mold spores are inhaled. These are primarily produced by trees, grass and ragweed. Repeated exposure to allergens will cause the production of antibodies which accumulate on the cells of your nose, eyes or bronchial tubes. When enough sensitizing antibodies have been formed, then contact with pollen or mold spores produces symptoms of hay fever. There is a release of histamine, a substance which causes swelling of the mucous membranes of the nose and eyes. Other substances cause the smooth muscle of the bronchial tubes to go into spasm. The reactions occur.

Clean Breathing Apparatus With Enzymes. You need to

clean away the accumulated allergens on your trillions of respiratory and related breathing cells. By "scrubbing" them away, enzymes are then able to perform another vital anti-hay fever function — rebuilding the fragile, eroded walls of your bronchial tubes. It is this *double* action that will give you natural resistance to hay fever.

Fasting Program: At the approach of the hay fever season in your region, plan to go on a unique but amazingly beneficial fasting program. On *alternate* days of the week, enjoy raw foods. A suggested method:

Day One: usual meal program.
Day Two: fresh raw fruits and their juices exclusively.
Day Three: usual meal program.
Day Four: fresh raw vegetables and their juices exclusively.
Day Five: usual meal program.
Day Six: fresh raw fruits and their juices exclusively.
Day Seven: usual meal program. *Optional:* raw fruit fast in the morning; raw vegetable fast in the evening.

Benefits: Without interference or "competition" from other foods, your digestive system is now able to fully metabolize these raw foods, use the enzymes for saturation-repair of the fragile lungs and then build resistance to pollen sensitivity.

In particular, on this alternate day fast, enzymes are able to *neutralize* the sensitivity-causing histamine so that there is a lessening of the nasal swelling that provokes an uncomfortable series of reactions.

Brings Relief in One Week

Margie O'Q. would dread the approach of warm weather. It meant the release of pollen into the air and her dreaded hay fever which became worse with each year. Unable to take medications because they made her drowsy and dizzy, she wanted a natural healer. She spoke to an allergist who said that she needed to boost the power of self-resistance through stronger breathing channels. This could be done through enzyme-rebuilding of the sensitive cells and tissues. She was told to follow the simple alternate day fasting program as described above. Margie O'Q. did so. Within three days, she could breathe deeply with hardly any sneezing fit. In one week, she could venture forth in a pollen-filled field and breathe to her heart's content without any symptoms. She had used enzymes to free herself from hay fever!

HOW TO BREATHE SAFELY IN AN
ALLERGEN-FILLED ENVIRONMENT

Other Common Allergens. Many folks go into sneezing and coughing spasms when they come in contact with common allergens. While they do not have asthma or hay fever, they do have a respiratory sensitivity to such allergens as: pollens, molds, house dust, animal danders (skin shed by dogs, cats, horses, rabbits), feathers (as in feather pillows), kapok, wool, dyes, chemicals used in industry, medicines, insect sprays and stings, too.

Reactions That Occur. When this common allergen is absorbed into the bloodstream, it stimulates certain small white blood cells (called *lymphocytes*) to produce special substances known as allergic antibodies. These will react with the allergen and produce allergic inflammation, and irritation. In particular, vulnerable areas are the nose, eyes, lungs and digestive system. These reactions may occur on occasion or frequently, depending upon the level of exposure and the degree of capillary fragility.

Rebuild Collagen With Enzyme Catalyst Foods

To Protect Yourself: Rebuild your respiratory collagen with the use of high enzyme catalyst foods. Each day, plan to have a big platter of fresh raw fruits and fresh raw vegetables. In so doing, you will be sending a high concentration of enzymes that will rebuild the protein substance known as collagen. This life-giving substance will nourish and repair your breathing components. This builds resistance to sensitive substances.

Dilute Lymphocytes. When you drink a glass or two of any enzyme catalyst raw juice, you send these miracle workers directly into your bloodstream. Almost immediately, the enzymes will create a natural dilution of your lymphocytes, reducing their effectiveness. The enzymes will create a "natural control" on the release of allergic antibodies which are the source of irritation. This type of "blockage" will help build an "internal fortress" that will shield your body from the provoking substances that would otherwise create an allergic attack.

Allergy-Relief Enzyme Tonic. Combine equal amounts of raw cabbage juice, carrot juice, cucumber juice, tomato juice. Add a bit of lemon or lime juice for a tangy taste. Stir vigorously. Sip slowly.

Benefit: The powerhouse of enzymes in this tonic act as instant catalysts. They use the Vitamin A and C of the vegetables, together with the magnesium and manganese of the cucumber to create the formation of collagen. Within moments, the enzymes catalyze the collagen repair of your breathing apparatus. You'll soon discover you can enjoy a breath of air (and more) without distress. It's the natural way to enjoy relief from allergic symptoms.

Enjoys Allergy-Free Breathing

Bud S.G. could look back on a lifetime of super-sensitivity to ordinary house dust or flecks from a favorite animal. He would sputter and sneeze. His eyes would water. Medications eased but did not erase his distress. A visiting lecturer on television suggested that such folks try a fresh glass or two of a raw vegetable juice in the morning, another at noon, a third in the evening. This triple-fortification would provide "around the clock" enzyme-collagen repair of the sensitive lungs and bronchial tubes. It would help the process of breathing. Bud S.G. tried it. Within four days, he could breathe more easily, even when his wife kicked up a dust storm in her spring cleaning. Within eight days, he could enjoy "total freedom" from sensitivity, whether house dust or animals. He drinks just one glass of a raw vegetable juice daily to maintain his freedom from allergies.

GOOD NUTRITION WILL EASE
ASTHMATIC-ALLERGIC DISTRESS

"Good nutrition should help make life easier for the asthmatic," says Marshall C. Goldberg, M.D., of the American Academy of Pediatrics and the American Academy of Allergy. Dr. Goldberg offers these guidelines:[1]

Vitamin C May Help. Most asthmatics have too low a level of Vitamin C (used by enzymes) in their bodies. "Vitamin C is quickly used up in the asthmatic to detoxify toxic materials entering the body. There are tests that show when 300 milligrams of Vitamin C were given at 15-minute intervals to a group of asthma patients during an attack, some experienced immediate relief and the remaining folks experienced relief within one hour," he reports. (*Note:* Raw enzyme catalyst fruits are powerhouses of Vitamin C.)

---

[1]*Parker Natural Health Bulletin,* West Nyack, New York 10994. Vol. 8, No. 17, August 14, 1978. Available by subscription.

Good Nutrition Can Ward Off Toxins. Vitamins may not only ward off allergens but also discourage the entry of harmful substances into the cells. A nutritional (enzyme plus Vitamin C) deficiency renders the cells *permeable,* which means they are prone to absorb toxic substances. Dr. Goldberg explains, "Imagine the nutrient-poor cell as a fine strainer in which the holes are enlarged. It works both ways. Nutrients can escape the permeable cell while toxins and foreign substances can enter."

Needed: Vitamins A and E. Dr. Goldberg also says that Vitamin A is essential to tissue integrity and that Vitamin E is another key element. (A yellow vegetable juice with a bit of wheat germ or bran, blenderized, gives you this enzyme catalyst combination.) "All the nutrients are necessary to keep foreign substances from entering the cell," he adds.

Fluid Balance Important. Asthmatics have occasional salt cravings that may indicate a nutritional deficiency. Dr. Goldberg says that "artificial or junk foods and refined foods are not easily eliminated and cause bacteria buildup. This depletes the salt and other elements in an almost vain battle to prevent adrenal exhaustion. Because the asthmatic's body is low in salt, the prescribed drinking of much water and clear liquids during an attack helps reestablish the balance of body fluids upset by salt depletion. Heavy liquid intake helps to reduce the inflammation by plumping up the tightened bronchia," Dr. Goldberg explains.

Enzyme Catalyst Action to the Rescue. From the preceding, it is apparent that a glass or two of fresh fruit or vegetable juices provides speedy enzyme catalyst action to help rescue your lungs from an allergic attack. These enzyme catalysts soothe the parched membranes of the lungs, promote collagen healing, thereby ease inflammation and "plump" up the constricted bronchia. Relief is just an enzyme catalyst swallow away!

RHYTHMIC BREATHING IS REFRESHING

Yoga enthusiasts call rhythmic breathing "inhalation therapy." Physicians call it "aerobics." Allergists say that it offers a breath of fresh air to the parched organs. Nutritionists recognize that *enzymes have stronger catalyst repair powers when sufficiently oxygenated.* Therefore, a simple program of rhythmic breathing will put unique power of allergy-healing into your enzymes.

Benefits of Rhythmic Breathing. With each complete inhalation and exhalation, carbon dioxide is burned and waste material is eliminated from the blood. This makes it more effective for enzymes to be transported to the vital body organs for healing and regeneration. Rhythmic breathing also helps supply your blood with oxygen, which is needed by every body cell. It changes dark venous blood into fresh red enzyme-carrying blood. Your entire body is rewarded with a feeling of refreshing vitality.

Easy-To-Follow Exercise. Wearing loose garments, lie down on your bed or a towel-covered floor. Breathe in deeply. Hold it for five seconds. Breathe out until your entire respiratory system feels cleansed. Repeat this allergy-cleansing exercise ten times in the morning and ten times at night. It helps to oxygenate your enzymes and makes you feel refreshed with a welcome "breath of life."

PROTECT ENZYMES AGAINST
SUDDEN TEMPERATURE CHANGES

These may cause enzymatic power to become "shocked" or even destroyed. This may let your defenses down and allergies may strike. Some common sense approaches will be enzyme-satisfying.

To Begin: There are times when, for no apparent reason, your nose runs, your eyes fill with tears and you start to sneeze. This condition may last from five to thirty minutes. It may recur throughout the day.

Possible Cause: Sudden temperature changes may cause your respiratory nasopharynx nerves to constrict, "choke" enzymes and trigger off a reaction known as *vasomotor rhinitis.*

Enzyme-Soothing Suggestions. Avoid going from one temperature extreme to another. For example, if your office is overheated (or the air conditioner is on full blast), spend a few minutes in a temperature lobby or hallway before leaving the building. Avoid touching very cold surfaces with your hands or feet. Don't put your warm feet on a cold floor before getting in or out of your bed.

By avoiding these sudden temperature changes, you'll keep your enzymes in a contented condition and they will be better able to perform their catalyst healing tasks throughout your body. You'll also protect yourself against sudden allergic attacks caused by these temperature shocks.

HOW TO HELP EASE ALLERGIC SYMPTOMS

Here is a set of 10 steps that help you cope with and ease allergic distress. They are aimed at bringing about a feeling of relaxed contentment. This is the ideal environment for enzyme catalysts to create anti-allergy collagen for your respiratory tract. So you can relieve, avoid and "insulate" yourself against severity of allergic symptoms with these enzyme-soothing steps:

1. Do not drink alcohol. It tends to dilate blood vessels (as do allergic substances) and may precipitate an attack.
2. Try not to have any arguments or become involved in stressful situations. Emotional tension is enzyme-destructive and is believed to set off some attacks.
3. Do not overexert yourself or become fatigued. Do not make drastic temperature changes from one room to another. For example, you should not go from a cool indoors to a hot outdoors (or vice versa) without waiting in a neutral area to allow your enzymes to adjust to the change.
4. Do not eat or drink anything that is icy cold or extremely hot. Do not eat heavily spiced foods or beverages.
5. Give up smoking. It chokes your respiratory organs, actually "burns" your enzymes and makes you more vulnerable to worsening allergic symptoms.
6. Do not swim in a chlorinated pool. If you're swimming in an ocean or lake, do not suddenly plunge into the cold water. Ease yourself in slowly.
7. Listen to local weather forecasts or call your local health department for the daily pollen count. If it is high, do not open windows. Stay indoors in an air-conditioned environment as much as possible. But take care that the room does not become too cool. A comfortable temperature is about 75°F.
8. Avoid using powdered soaps and insecticides.
9. Do not cut flowers or do any gardening or household painting.
10. Do not go for a ride in the country, for you may find yourself in the midst of a pollen area and suffer an attack.

Basic Guideline: Do *not* abuse your enzyme catalyst system. By treating this system and your body with gentle care, you can help build immunity *from within* so that you can enjoy freedom from allergies from the outside!

IN REVIEW:

1. Bioflavonoid enzymes in citrus and other fruits will help build strong resistance to and freedom from severe asthma.

2. Within nine days, Ned T.W. was able to say "goodbye" to life-long asthma on a simple bioflavonoid program.

3. Increase intake of fresh fruits which are super-powerful in bioflavonoid enzyme content.

4. To overcome hay fever, try the enzyme fasting program. Margie O'Q. was able to enjoy relief in just one week.

5. Breathe easily in an allergy-filled environment on an enzyme catalyst program that dilutes lymphocytes and eases symptoms ...almost at once.

6. Bud S.G. overcame sensitivity to house dust on a raw vegetable juice program.

7. Good nutrition, notes an allergist, eases asthmatic-allergic distress. Emphasis is on vitamins needed by enzymes for respiratory rejuvenation.

8. Refresh your breathing with thythmic exercises.

9. Be kind to your enzymes on the simple 10-step allergy-easing program.

CHAPTER 14

Sprouts: The Most Powerful
Enzyme-Catalyst Food You
Can Grow at Home

Right in your own home, on top of a kitchen table, on a window sill, or any spare corner, you can grow one of the most powerful enzyme-catalyst foods that are known to exist. These are called *sprouts.* Here is one single type of food that is almost perfect in its content of a variety of nearly all known nutrients. But its most health-strengthening ingredient is that of its super-enzymes. When you sprout seeds, you double, triple, even *quadruple* the potency of enzymes that would exist either in the seeds or in the food that is grown from those seeds. If you are seeking just one food source of dynamite enzymes...it would be seen sprouts!

WHY SPROUTS ARE SUPER-ENZYME COLLAGEN FOODS

To Begin: Sprouted seeds are vital to you because they are a prime treasure of fresh nutrients and enzymes already placed there in collagen-building balance by Nature.

Vitamins: Sprouts are the best source of vitamins needed by enzymes. These include Vitamins A, B-complex, C, D, E, K and other micro-elements that become dynamically magnified and supercharged during the sprouting process.

Minerals: Most known minerals and trace elements exist in seeds. They become enhanced and increased in size and (most im-

191

portant) enzyme effectiveness during the brief and easy sprouting process.

Protein: Sprouted seed protein is living, growing and regenerative. It is ready for use by enzymes to build and rebuild your body almost immediately after consumption.

Important Fat. Sprouted seeds are very rich in *beneficial* fat (unsaturated) which is cholesterol free and used by enzymes for transport of vitamins and minerals to your trillions of cells and tissues. Enzymes will use the sprouted seed fats to keep your skin and hair from drying, your body properly moisturized, and your vital organs in healthful condition.

Special Hormones: Sprouted seeds are enzyme catalysts for body rejuvenation because they contain special hormones known as *auxins.* Enzymes immediately use these plant hormones to speed up the vital processes in your body cells and boost the youthful vigor of your endocrine glands. Almost immediately, the enzymes will use sprout *auxins* for this youthful hormone release and regeneration of your vital organs.

Unique Amino Acid Release. When you activate the seed by soaking it during the sprouting process, the biological machinery is put into action to activate all the life forces locked within the seed. In particular, sprouting tends to boost amino acid manufacture, with emphasis upon lysine and tryptophan. The nutritional value of the amino acids becomes enormously boosted during the process of germination (sprouting). Now, your enzymes can use these amino acids to create internal (and external) rejuvenation.

REBUILD BODY CELLS WITH ENZYME-COLLAGEN SPROUTS

Boost the power of self-rejuvenation with the use of fresh sprouts that are a treasure of enzyme-collagen. Sprouting can, for example, increase the Vitamin C content of soybeans by as much as 500%. The B-complex of sprouted oats is some 13 times higher than the original product. Other nutrients are similarly magnified. This means your enzymes have *more working material* available for cell rebuilding.

Helps Control Cellular Aging. This condition is caused by cellular breakage or so-called mutations. Certain toxic wastes in the system tend to cause cellular changes as well as exert an inhibitory effect by interfering with enzymes. To protect or control this risk to cell life, *powerful* enzymes are needed. Sprouts are powerhouses of enzymes as well as the vitamins and minerals and other nutrients described above. They have the dynamic ability to actually "knock out" interference and "destroy" harmful elements. Now enzymes can use nutrients for cellular regeneration without interference. This helps enzymes control and even halt cellular aging. Small wonder that sprouts are considered the enzyme solution to unnecessary aging!

WHAT TO SPROUT

Examples of seeds that are easy to sprout within six days are alfalfa, mung beans, soy beans, garbanzos (chick peas), garden peas, mustard seeds, cress, black and red radish, purslane, chia, flax, broad beans, unhulled sesame seeds, safflower, sunflower seeds, grains. Also try lentils, dried peas, millet, barley, buckwheat, fava, lima, pinto beans, wheat berries.

HOW TO BUY SEEDS

Most of these seeds and grains and beans are available in your supermarket or garden-supply stores. You may find still others at a local health store. Do *not* buy seeds that have been chemically treated. If in doubt, seek out an organic seed farmer for a good source of sprouts.

HOW TO SPROUT

You Will Need:

1. Untreated seeds as described above. Buy only a few ounces to start with.
2. Clean glass jar that holds about one quart.
3. Clean square of cloth or paper towel to cover jar opening, and elastic band to fasten.
4. Strainer or sieve for rinsing and draining your sprouts.

Steps To Follow:

1. *Sort and wash.* Start with about 1 tablespoon of seeds or 1/2 cup beans or grains. Rinse the seeds well with clean lukewarm water. Pick out and discard shriveled or broken seeds.

2. *Soak overnight.* Cover the seeds with plenty of clean lukewarm water. Then set aside for an overnight soaking.

3. *Drain and rinse.* In the morning, drain the seeds and rinse at least twice in clean lukewarm water. After the last rinse, drain the seeds well and return to the jar. Seeds should be moist, but they should not be wet.

4. *Keep dark and moist.* Cover the jar with the cloth and secure with an elastic band. Turn the jar on its side and rotate so that the seeds are in a thin layer. Place the jar, on its side, in a dark closet or cupboard.

5. *Rinse at least twice daily.* Thoroughly rinse the seeds with clean lukewarm water. Each time drain well and return to the jar as before. By the second day, sprouts should begin to appear. Small seeds usually take less time than beans. Sprouting is faster in warm weather. *NOTE:* Thorough rinsings are essential to prevent mold growth and "souring" of the sprouts. Soy beans tend to mold more easily than other beans and seeds. It is a good idea to rinse them more frequently, five or six times daily, and to use clean *cool* water.

6. *When sprouts are grown.* Sprouts can be grown to various lengths. But the favorite lengths for some sprouts are:

Alfalfa   — one to two inches
Lentils   — one inch
Mung beans — one and one-half to two and one-half inches
Soy beans   — one to two inches
Wheat   — one-fourth inch

Aim for one to two inches for most seeds.

Note: If you like green sprouts, place the jar in sunlight for 30 to 60 minutes, just before harvesting.

7. *How To Harvest Home-Grown Sprouts.* Remove from the jar when sprouts have reached their desired length. (This is called harvesting.) Then refrigerate. Just place them in your refrigerator in a *closed* container to preserve their freshness and nutrient value. NOTE: Plan to use within three days for maximum enzyme-catalyst and nourishing power.

HOW TO ENJOY YOUR SPROUTS

They pack an enzyme "wallop" if you can eat them right away. The enzyme catalyst power is at its peak. Within moments, enzymes will begin the collagen regeneration of your entire body. Try to eat them as soon as possible. They give you a sweet crunch, when chewing, and a unique flavor that you will always enjoy. You can enjoy them raw or cooked. Of course, if cooked, there is a slight loss of enzyme power. Still, you may want to try them any of these ways for variety.

Raw Sprouts. A bowl of these sprouts will supercharge your body with enzyme catalysts and start you on the road to total rejuvenation. Try eating these crunchy, chewy-good sprouts as they are; or, include in salads or sandwiches, or eat as a finger food or a garnish for soups. Mix into green salads. Combine with your favorite whole grain breakfast cereal.

Lightly Cooked Sprouts. Use as a vegetable alone or in combination with other vegetables. Sprouts may be added to soups, stews, casseroles, eggs and Oriental dishes. Cook sprouts only long enough to remove the "raw taste." Saute in a small amount of liquid vegetable oil. Or steam in a small amount of boiling water.

SPROUT RECIPES

Sprout Saute: Brown minced onion in hot vegetable oil. Toss in raw mung bean sprouts or lightly cooked soy sprouts. Season with favorite herbs. Stir until well blended and heated through. Serve promptly.

Scrambled Eggs With Sprouts: Combine raw mung beans sprouts or slightly cooked soy bean sprouts with slightly beaten eggs. Add herbs, onions or other flavorings as desired. Scramble in a lightly oiled skillet.

Alfalfa Sprout Salad: In a large bowl, combine 1 cup chopped celery, 2 cups grated raw carrots, 1/2 cup sun-dried raisins and 1 cup raw alfalfa sprouts. Toss with a mayonnaise dressing. Serve chilled.

Grilled Cheese With Sprouts: For each serving, place a slice of whole wheat bread under the broiler and toast lightly. Remove and add a slice of cheese. Return to the broiler until melted. Top with a tomato slice and a generous sprinkling of alfalfa sprouts.

Cucumber Salad With Sprouts: Combine 1/2 cup vegetable oil, 2 tablespoons of apple cider vinegar, 1/2 teaspoon dill weed, teaspoon of salt substitute or kelp. Pour over a thinly sliced cucumber and refrigerate for two or more hours. Just before serving, arrange salad greens, the cucumber slices, tomato wedges and raw sprouts (alfalfa or mung) on a serving platter. Sprinkle with the remaining dressing.

Rejuvenates Digestive System in Three Days

Vivian DeP. complained of "sour stomach." Much of what she ate either felt like "dead weight" or else made her stomach rumble and churn. She was troubled with stomachaches almost daily. Patent medicines only worsened her condition. She had to do something to ease her digestive trouble or else she would stop eating and just starve herself to death! A local nutritionist suggested she boost her enzyme powers. She had to put "chewing power" into her digestive system. The nutritionist advised her to eat at least three cups of raw sprouts each day. She could enjoy her regular meals but each one *must* contain the sprouts. Vivian DeP. tried it. Within one day, her stomach stopped grumbling. At the end of three days, she was totally relieved of stomach upset. She threw away her patent medicines. Now, if she has any distress (occasionally), she just has a cup of raw sprouts and she is healthy again.

Benefits Of Raw Sprouts. The very act of sprouting amplifies and quadruples the content of enzymes that were in the original seed. These enzymes neutralize the excess stomach acidity and also dilute the toxic wastes. The enzymes then create a catalytic response to wash out these digestion-interfering substances. In a matter of hours, the sprout enzymes have cleansed and regenerated the digestive system. Because sprouts are *potent concentrates* of enzymes, they become the most powerful sources of these digestion and body strengthening foods known. If you must have only one enzyme food, let it be sprouts!

HOW A "SPROUT FAST" CAN REJUVENATE YOUR BODY OVERNIGHT

Nip the aging process in the bud with sprouts! They introduce a powerhouse of live enzymes that spark your digestive organs to metabolize the nutrients of the food you eat into substances that can rebuild your body from head to toe...*even while you sleep!*

From 70 To Under 40 Within One Night

Susan I. was troubled with aging skin, loss of vitality, feeling cold hands and feet (even in warm weather) and general tiredness. She felt the onset of old age even though she was in her middle years.

When someone offhandedly whispered that she looked 70, Susan I. decided to do something about the problem. She went to a dietician who made a simple blood test. Susan I. needed more enzymes to wake up her sluggish metabolism. The dietician recommended that she go on an easy one day "sprout fast."

Simple Program: For one day, your meals should consist exclusively of freshly prepared sprouts. You may take fresh fruit or vegetable juices throughout the day but *not* with the sprout meals. The reason here is that the liquid could "wash away" the power of the enzymes and nutrients introduced by the digested sprouts. *No* cooked (or other) foods for this one day.

Wakes Up Years Younger

Overnight, the sprouts worked their rejuvenation powers. Next morning, Susan I. awakened with a feeling of limitless energy. She felt warmth flowing through her rejuvenated body. Her skin was firm, her movements alert. Now, she no longer looked 70, but with a twinkle in her bright eyes, said she was under 40.

Enzyme Miracle. This rejuvenation process had occurred *overnight!* That was all the time required by the supercharged sprouts to create their internal and external makeover process. No other enzyme food can equal this power!

Enjoy Renewed Youth With Enzyme-Catalyst Sprouts

When you chew and swallow enzyme-catalyst sprouts, almost immediately, these cell-tissue rejuvenators are at work in building and rebuilding your entire body.

Your vital organs, your various systems, your nerve and circulatory networks, your bloodstream will all become revitalized through the use of sprouts. You may enjoy continual rejuvenation by eating sprouts at least once a day. For "emergency" help, go on a simple one day "sprout fast" and let the enzyme-catalyst power rejuvenate your body while you sleep! You'll wake up looking and feeling years younger!

IN A NUTSHELL:

1. The most highly concentrated source of enzymes and other nutrients is that of seed sprouts. Chewy good and tasteful, they are speedily effective in cell-tissue regeneration. Work while you sleep.

2. Grow your own sprouts in a jar, right at home.

3. Enjoy your sprouts as they are or in any of the methods described.

4. Vivian DeP. rejuvenated her digestive system in three days with the enjoyment of enzyme-catalyst sprouts.

5. Susan I. went from 70 to under 40 within one night on a "sprout fast." The years just rolled away while she slept!

6. Enjoy renewed youth (very quickly) with the tasty joy of enzyme-catalyst sprouts.

CHAPTER **15**

Sources of Power Enzyme-Catalysts

In Everyday Foods

To keep your body in tiptop share, you need a daily supply of enzyme-catalyst foods. Enzymes play the supreme role in cell-tissue regeneration and all life's processes from head to toe. Virtually all foods you eat, however nourishing, remain indigestible until enzymes work on them, break them down into simpler substances, and then help them become absorbed in your bloodstream to build and rebuild your vital organs.

The Youth-Building Power of Enzymes. In a twinkling, enzymes perform biological transformations that are either difficult or impossible to perform in a laboratory. *Example:* To digest any animal source food would require hours and hours of chemical onslaught under laboratory conditions. Yet a strong enzyme supply in your digestive system can create this reaction within a few hours, and then send the metabolized nutrients via your bloodstream to all body parts for nourishing. There can be no substitute for enzyme-catalysts to keep you alive and healthy.

HOW ENZYMES KEEP YOU IN YOUTHFUL HEALTH

When you have *quantity* plus *quality* enzymes, they can perform miracles in terms of youthful health.

Examples: Some enzymes are oxidants (fuel burning). They take a piece of food, a tiny fragment, and start it on a series of biological reactions that will create the miracle health foundation of all life: *adenosine triphosphate,* or ATP, for short.

Enzymes Create ATP. How essential is this dynamic substance that is created ONLY by enzymes? Basically, ATP is a tiny storage battery that is essential for the retention and then transfer of energy to your trillions of living cells. Enzymes cause ATP (the fundamental component of all living matter) to transfer one of its three phosphate groups to other compounds, together with a collection of immediately available energy substances for use in biological reactions. The basic one is cell-tissue regeneration and renewal. Without ATP, there would be no life. And...without enzymes, there would be no ATP. So enzymes are the spark of life.

Enzymes will catalyze ATP to release stored energy to make your muscle fibers contract. Every time your heart beats, every time your eyelid blinks, every time you move your hand to turn the page of this book, every time your eyes move from one line to another, enzymes are required to stimulate ATP to give you energy to perform these functions.

Improves Brain-Nerve Health. The same reaction occurs in your brain and your nervous system. Enzymes produce a substance called acetylcholine; they catalyze the substances required to use this nutrient to help transmit messages across nerve junctions. In effect, enzymes create acetylcholine to help you think better! An enzyme deficiency would mean an acetylcholine deficiency and this could run the risk of brain fatigue or, worse, senility! You need a daily supply of available enzymes in order to create this brain-nerve reaction and just about every other response that gives you life. Enzymes come to the rescue to give you the look and feel of healthy living.

Daily Supply Of Enzymes Is Vital

Because enzymes are destroyed during the life processes, they need to be replaced on a daily level. Remember, too, that enzymes are highly selective. A given enzyme will catalyze one kind of reaction but no other. There can be no substitute. If you lack several fat-melting enzymes, then you run the risk of an overload of this nutrient. Other enzymes cannot do the task for those that are missing. Also, enzymes are sensitive to changes of digestive acidity, changes of temperature. All enzymes are destroyed by heat. A fever will deplete much of your enzyme supply. Cooking food at a temperature higher than 120°F. will also destroy enzymes. So your goal is to get a daily supply of enzymes from raw or living foods.

YOUR ENZYME FOOD SOURCE GUIDE

Here is a listing of a wide selection of foods that are prime sources of enzymes. You should plan to eat a *variety* of these foods to assure your body a variety of available enzymes. In so doing, you will be boosting your supply of these catalysts and helping to build and rebuild health from within. You'll then be able to glow in youthful health from your outside!

APPLES. Enzymes will use the natural pectin for rebuilding of your cells and tissues. Produce a cleansing of toxic accumulations.

AVOCADO. The vegetable protein content and the high concentration of vitamins and minerals are used by enzymes for body invigoration, repair and regeneration.

BANANAS. Helpful for digestive improvement and also for improved carbohydrate metabolism. Help increase energy.

BERRIES. Contain a natural citric and malic acid content that is used by enzymes to promote improved blood building.

CRANBERRIES. Enzymes will use the natural fruit acid supply to create a detoxification program within your body.

CHERRIES. High in natural fruit sugars, they produce swift energy.

DATES. An excellent source of simple fruit sugars, dates also contain a special type of fruit protein that is used by enzymes for internal regeneration.

FIGS. A high mineral content that is similar to human milk; enzymes will use it to make more calcium available to your body and nervous system, especially for your skeletal structure.

GRAPES. Considered the "king of fruits," a good source of iron and vitamins. Enzymes use them for blood and body rebuilding and also for cleansing of the cells and tissues.

RAISINS. Considered dried grapes, they are highly concentrated sources of energy. Enzymes, themselves, are energized by raisins and use stored up energy to give you more vitality in body and mind.

MANGOES. A tropical fruit, it appears to have vitamins and minerals that are used by enzymes for basic improvement of health.

MELONS. All are considered good enzyme foods. Enzymes from melons are able to digest very "tough" meats. Just plan to eat a fresh melon platter after a "heavy" meal and it will digest more comfortably and speedily.

WATERMELON. A good source of Vitamin A that is used by enzymes for rebuilding of skin and vital organs. A good source of easily assimilated enzymes because of the high natural liquid content.

NECTARINE. Very potent enzymes that are able to digest and metabolize strong protein foods from an animal source.

ORANGES. A high concentration of vitamins and minerals, especially Vitamin C and potassium. The enzymes will use these nutrients to rebuild your trillions of cells and tissues and also improve the health of your bloodstream. Guards against cell fragility.

PAPAYAS. Considered a "tree melon" because of its resemblance to that fruit. A high source of vitamins and very powerful fat-starch dissolving enzymes.

PEACHES. Contain a unique type of "fruit ether" which tends to soothe the organism; enzymes will create a feeling of digestive contentment during the rebuilding process.

PEARS. Similar to the apple (there is a botanical relation), pears contain a high natural sugar content that seems to boost efficiency of enzymes. Pears may be considered a good source of energy enzymes.

PLUMS. Juicy good, they contain a good supply of vitamins and minerals of a somewhat energetic source. They invigorate enzymes and accelerate the digestive-metabolic processes.

PRUNES. A dried plum, this fruit is a high concentration of natural sugars and super-energetic enzymes. A few prunes eaten after a meal will promote more youthful digestion and body energy. The prune also contains unique enzymes that assist in toxic cleansing and, most important, natural intestinal regularity.

TOMATOES. A good source of natural acids and alkalines that tend to "scrub" out wastes from the system and prepare them for easier elimination. Tomato enzymes can perform this cleansing action; they also use tomato nutrients to improve the acid-alkaline balance of the bloodstream.

ACORNS. These nuts, from the oak tree, are considered farinaceous. When properly cracked and peeled, the acorn meat is a high concentration of vegetable protein and unique enzymes that are able to provide almost total nourishment.

ALMONDS. A high phosphorus content along with calcium in this food make it very desirable for basic blood, bone and nerve strengthening. Almond enzymes can create this "structural rebuilding" of your body.

BRAZIL NUTS. A good source of polyunsaturated fatty acids as well as calcium and magnesium. Nut enzymes will use these nutrients to protect against cholesterol overload. Enzymes will also use the minerals for basic organ rebuilding and bone strengthening.

CASHEW. A high concentrated source of plant amino acids and enzymes that work to enrich your bloodstream, revitalize your essential organs, and promote overall rejuvenation.

CHESTNUT. Contains an abundance of nut carbohydrates as well as enzymes that work to speed up the energy-producing system.

COCOANUT. Cocoanut milk contains a high amount of iron, phosphorus and potassium along with important vegetable fats that are enzyme metabolized and used to rebuild your body.

HICKORY NUT. The protein of this nut is of a very high order. Because of its thick shell, it is not vulnerable to outside exposure and the enzyme content is especially potent. When you crack the hickory nut, you give yourself an almost "perfect" protein food with the most powerful enzymes you can have from such a plant. A plate of shelled hickory nuts can make up a delicious meal.

PECAN. Contains natural polyunsaturated fats that are essential to help protect against cholesterol buildup. Also a prime source of vitamins and minerals and special "oily" enzymes. That is, they are able to lubricate your internal organs so you function more smoothly and efficiently.

PIGNOLIA or PINE NUT. A high protein source, along with valuable polyunsaturated fats and enzymes. This makes up a good health package.

PISTACHIO. Greenish in color. The greener they are, the richer in enzyme and nutrient content. They are high in protein and the oil is easily digested. The pistachio contains little indigestible cellulose (roughage) so it is nearly all edible food. Enzymes are very active and can help revitalize the entire body.

WALNUT. Improve your balance of polyunsaturated fatty acids with this variety. It contains enzymes that are able to promote better cleansing of accumulated wastes and build a healthier body.

PEANUT. Ranked high in enzyme biological value because of its unusually high protein quantity and quality. It can be used as a meat substitute because it is almost a complete food. Enzymes from peanuts are very concentrated and able to create strong revitalization for the entire body.

BEETS. Grated raw and added to a vegetable salad, they form a colorful and delicious addition. A prime source of juicy enzymes which work to enrich your bloodstream and help improve overall metabolism.

BROCCOLI. Use chopped in a salad or as a briefly cooked vegetable. If you steam for a few moments, the enzyme content is slightly diminished but of good value, nevertheless. Helps provide enzyme-propelled minerals that will invigorate your body and improve basic health.

CABBAGE. Contains factors that appear to improve the digestive organs. Enzymes will use these unusual substances for rebuilding of the digestive tract and guard against lacerations and wounds that may erupt as ulcers.

CARROTS. A high form of carotene which is used by enzymes to be transformed into Vitamin A for nourishing your eyes, your skin and your internal organs.

CAULIFLOWER. Chop very fine and mix with salads. This vegetable is a good source of alkaline minerals that are used by enzymes to help maintain healthy cells and tissues.

CELERY. An excellent source of sun-improved enzymes together with a high treasure of minerals that will improve your bloodstream and provide important roughage for digestive-intestinal cleansing. Enzymes use its cellulose fibers to act as an "intestinal broom" and sweep away wastes and sediment that might otherwise cause erosion of your vital organs.

CHIVES. Small, slender leaves should be chopped fine and used as a flavoring for salads. They contain potent electrolytes and trace elements which are used by the enzymes to boost your body health.

CUCUMBER. A prime mineral food, its enzymes use these nutrients to help improve the health of your skin and hair. Enzymes will use cucumber minerals to help regenerate your vital organs through cell-tissue rebuilding.

DANDELION GREENS. A powerhouse of high concentrations of enzymes that are able to create speedy regeneration. Powerhouse of vitamins and minerals that work with the enzymes to rejuvenate your vital organs.

ENDIVE. A hardy food, it is great for using in a vegetable

salad. It has a slightly bitter taste which attests to its unusually strong enzymes. Almost immediately after being thoroughly chewed, it speeds up your digestive process, is able to "attack" tough animal-source foods and improve upon assimilation of removed nutrients.

FENNEL. It has an aromatic taste and fragrance. Use the high-enzyme seeds and fragrant young leaves for a salad. Will help improve the quality of the bloodstream and your basic circulatory and assimilative systems.

GARLIC. Plan to eat a clove raw (chew well) or dice and add to a salad. Contains unusual enzymes that are able to help control blood pressure through cleansing away of toxic wastes. Garlic enzymes will use vitamins and minerals to help stabilize the nervous system. A natural internal cleanser.

GRAPEFRUIT. A prime source of important enzymes that help to digest very strong foods of the animal-source group. Especially potent when eaten after a heavy meat and/or fatty meal. Grapefruit enzymes will pierce the tough membrane of the meat and accelerate the transformation of protein into usable amino acids for body regeneration. Makes a healthful and juicy good dessert, after any meal.

KALE. A form of cabbage with foliage (used as greens) that contains high concentrations of enzymes. It will help boost production of body enzymes and improve overall digestive powers.

KOHLRABI. A root vegetable, its mineral content is especially highly concentrated. Its enzyme supply is very rich since soil nutrients cause a great proliferation during the growing process. Chew kohlrabi thoroughly; you may also chop fine and use in salads.

LETTUCE. The leaves and the woody stalk-like portions are succulent sources of enzymes. They will help improve your mineral metabolism and also revitalize your bloodstream. All types of lettuce should be used for their good taste and enzyme content.

MUSHROOMS. An edible fungus, this food is grown in moist, cool and dark places. A good source of roughage. Enzymes are comparatively mild here because of the tranquil environment in which mushrooms are grown. Yet they do exert a feeling of contentment when enjoyed with a meal. Fresh, raw sliced mushrooms, flavored with a bit of oil and a sprinkling of favorite herbs will add up to a goodly amount of body cleansing enzymes, not to mention the good taste.

MUSTARD GREENS. Resembling spinach, this plant is considered a "wild" plant because it proliferates freely in grain and pasture fields. This freedom enables enzymes to grow and multiply so that they are abundant and strong. Use mustard greens as part of a salad or for munching alone. They give you a high amount of mineral-enriched enzymes.

ONIONS. Yes, you *can* eat raw onions as part of any salad. The very volatile and pungent oils are sources of equally strong enzymes. Onion enzymes are important for cleansing the bloodstream of accumulated wastes; hence the reputation of this enzyme source as a blood builder. Onion enzymes are also able to influence the number of red blood cells and content of hemoglobin of the bloodstream. Plan to eat sliced or diced or chopped onions as a salad garnish every day. Many different onions are available with variations in taste but with similar enzyme potencies.

PARSNIP. The long, thick, white, sweet root grows deep underground where it absorbs the rich minerals from the soils. This creates a greater development of natural enzymes which tend to multiply in the depths of the cool mineralized earth. When ripened, parsnip enzymes are extremely vigorous in being able to pierce stronger foods and also to dislodge stubborn accumulations of wastes in the body. Parsnips, sliced thin and eaten as part of a salad, do offer a good internal cleansing and detoxifying that is holistically beneficial.

PEPPERS. The green variety will offer milder enzymes that work with minerals to help maintain better water balance in your digestive-eliminative systems. The red variety will offer a stronger form of "enzyme scrubbers" that will dislodge thicker accumulations of waste matter and prepare them for elimination. Green or red peppers are also high in Vitamin C which is used by enzymes for the regeneration of your trillions of body cells.

RADISH. This vegetable is a prime source of skin-hair nourishing sulphur. It also contains very strong enzymes (hence its "hot" taste, at times) that will take the sulphur and use it for improving the health and firmness of your skin; enzymes will also use the sulphur to nourish the follicles of your scalp and send forth hormones to help improve your hair. Plan to use radishes regularly for the enzyme-strong taste to balance out milder vegetables, if desired.

TANGERINE. A highly concentrated source of Vitamin C, this delectable orange-colored and sweet tasting fruit offers a fountain of juicy good enzymes. The high bioflavonoid content (found in the white fibers and "strings" of the segments) combines with Vitamin C and is powerhoused by the enzymes to rebuild collagen on your trillions of body cells. Plan to eat tangerines regularly for this enzyme-catalyst collagen regeneration of your entire body.

TURNIPS. The enlarged roots of this hardy plant offer a rich supply of soil-magnified enzymes. When you inhale the fragrance of this vegetable, you'll scent something unusually delectable. That is the high mineral content. The minerals are used by the concentration of powerful enzymes for improving the calibre of your bloodstream, strengthening your skeletal structure, and healing and revitalizing your body organs. Enjoy this high enzyme-catalyst food either sliced or diced as part of a salad…or with your favorite dressing, by itself. *Note:* The turnip leaves, when fresh, also make a high-enzyme mineral-rich salad. Use these as often as possible. *Tip:* Select turnips that are smooth and firm for top enzyme power.

BRUSSELS SPROUTS. These offer you enzymes that tend to regenerate the insulin making powers of the pancreas. They tend to create a more alkaline reaction and overcome that "burning" sensation also known as acidosis. Slice thin and use as part of a salad or, alone, with your favorite dressing.

SAUERKRAUT. This is a preparation of pickled cabbage. If salt-free, the fermentation process can accelerate and cause a proliferation of enzymes as well as minerals. (Avoid salted products which tend to be enzyme poor and also irritating to the mucous membranes of your system.) Salt-free sauerkraut can also cause an improvement in the natural enzymatic powers in the digestive system. Health stores and many supermarkets sell all-natural, salt-free sauerkraut. Enjoy your way to stronger digestive power with this high-enzyme power food.

WATERCRESS. The small but potent leaves of this flavorful food contain high concentrations of minerals *together* with enzymes that work synergistically. That is, one cooperates with the other, in order to create internal regeneration. Elements in watercress are also naturally acid-forming which means that these enzymes are able to act as a powerful intestinal cleanser. Watercress enzymes are also able to create healthful regeneration of the bloodstream, especially by increasing oxygen transmission for more youthful circulation. Enzymes will use watercress's concentration of phosphorus to create better utilization of carbohydrates, fats and proteins. This boosts repair of cells and production of energy, and also helps boost transference of nerve impulses. Enzymes are able to bring about improved glandular function through this better utilization of nutrients. Watercress should be enjoyed as often as possible for this catalyst regeneration reaction.

Feels "Reborn" With Enzyme Catalyst Foods

Henry D. felt his life "slipping" through his fingers, a little at a time. Each day became longer and longer. His work as a factory foreman became harder and harder. His fingers trembled. He had a sallow complexion. He shuffled with stooped shoulders. At times, he had difficulty understanding simple directives. He began to "lose his touch" in supervising others. He felt old before his time. He looked it, too, with a haggard expression, bloodshot eyes and furrowed skin. He felt he was finished.

Boost Raw Food Program. This was the advice offered by the factory's resident physician. He directed Henry D. to increase intake of an assortment of fresh raw fruits, vegetables, seeds, nuts. One day per week was to be devoted *exclusively* to raw foods and raw juices. Henry D. followed what he felt was "too simple" and "too good to be true" advice.

Skin Is Younger, Reflexes Sharper, Emotions Stronger

Within three days, Henry D. noticed a younger skin. His reflexes became more youthful. His emotional health was stronger. Now he had more strength and ambition than ever before. He felt he had been "reborn" through the enzyme-catalyst power foods he enjoyed daily. No longer did he feel he was "finished." Instead, thanks to enzymes, he felt he was "just starting" in life!

FIVE MIRACLE HEALING POWERS OF
ENZYME CATALYST FOODS

By partaking of a variety of the abovedescribed tasty raw foods (and their juices, whenever desired), you introduce a rich treasure of enzymes into your metabolic process. They exert their rejuvenating powers in five near-miracle healing reactions:

1. Raw food enzymes increase the micro-electric vitality in your cells and tissues, and thereby recharge your body with new youth.

2. Raw food enzymes work to stimulate improved cellular metabolism. This will help remake your cells and your entire body.
3. Raw food enzymes increase your cells' powers of resistance to aging. You are as young as your enzyme-nourished cells.
4. Raw food enzymes speed up the process of cellular renewal.
5. Raw food enzymes will help prevent toxic "suffocation" by cleansing your body of the accumulated wastes and erosive substances.

These are the foundation benefits of raw food enzymes. They set off a chain reaction that will promote the look and feel of youth, at any age.

Put power into your life with the use of everyday enzyme-catalyst foods. They taste every bit as good as they are effective in putting new energy and vitality into your entire body. You can, literally, eat your way to superior health with raw foods.

HIGHLIGHTS

1. Your body needs an ample quantity of high-quality enzymes from raw foods, every single day, in order to enjoy better health.
2. Enzyme-catalyze your body with the wide selection of listed foods. Enjoy them juicy fresh as often as you like.
3. Enzyme-catalyst foods helped "old" Henry D. feel "reborn" in just three days. His skin became young. His reflexes were sharp. His emotional health was strong. He was a *new* man in every sense of the word.
4. You reap the harvest of five miracle healing powers with the use of enzyme catalyst foods.

CHAPTER 16

Daily Guidelines for Your

Enzyme-Catalyst Health Program

It is simple and rewarding to plan your daily eating program around the use of enzyme-catalyst foods. You will discover a feeling of revitalized health by following these principles. Make these improvements in your lifestyle and you will be rewarded with better health, more youthful appearance and a feeling that life, indeed, is very beautiful...at any age.

1. EAT ONLY WHEN HUNGRY

Genuine hunger is a physiological need for food. It indicates that your digestive system desires food. This creates a more receptive *enzyme environment* that will boost catalyst powers. Try to establish regular eating times in order to put your body in healthy biological rhythm.

2. DO NOT EAT WHEN IN DISTRESS

If you have physical or mental pain, bodily discomfort of any sort, delay eating. These situations hinder the release of digestive enzymes and reduce powers of assimilation. It is healthier to skip a meal than eat when you are in the throes of upset of any sort.

3. EAT WHEN YOU HAVE LEISURE TO DIGEST

Enzymes work best if given time to catalyze after a meal. This bears up the old maxim, "A full stomach does not like to think." If you work (physically and/or mentally) either immediately before or after a meal, the foods will lie largely undigested (creating that heavy fullness) because tense enzymes are unable to work in a constricted environment. Healthy enzyme-catalyst action requires full attention of the system. Give yourself some rest before and after a meal so enzymes can work without strain or stress.

4. LIMIT DRINKING WITH MEALS

Liquids of any sort that are taken *with* meals tend to *dilute* and *weaken* enzyme-catalyst powers. Also, these liquids inhibit the flow of gastric juice. Your digestive system becomes "waterlogged" with too much liquid intake. Furthermore, drinking at meals leads to the bolting habit. Instead of thoroughly chewing your food (to release enzymes and generate digestive enzyme flow), you tend to gulp it down, or wash it down partially chewed. This reduces enzyme release. Plan to drink liquids about 60 minutes before and 60 minutes after a meal. If possible, lengthen the time span.

5. AVOID TOO HOT OR TOO COLD FOODS OR BEVERAGES

If too hot, there is a burning of the tissues of the stomach and a weakening of enzyme release. If too cold, enzyme action is halted as your system must wait until stomach temperature has returned to normal. A rule of thumb is simple: if the food or beverage burns your thumb, wait until the temperature is more soothing for your digestive system.

6. THOROUGHLY CHEW ALL FOODS

This is the most important guideline for enzyme-catalyst rejuvenation. Chewing has a double benefit: (1) salivary enzymes are activated to partially pre-digest food while still in your mouth and (2) digestive enzymes are then released so they can fully catalyze the chewed food when received in the stomach. Also, by thoroughly masticating your food, you eat in a more relaxed condition and this creates overall improved digestive health.

7. DRINK MILK? DO IT ALONE!

Not necessarily in hiding, that is. Rather, drink milk by itself and, preferably, *not* in combination with other foods. Milk acts as a gastric enzyme insulator. Its cream tends to limit the outpouring of digestive enzymes for some time after the meal is eaten. Milk is more fully digested in your duodenum (part of your digestive system) which means that in the presence of milk, your stomach does not fully respond with its enzyme secretion. This limit means that other foods taken with milk are not fully digested. *Note:* The use of acid-based fruits with milk is a satisfactory combination since enzymes are able to work efficiently upon both foods. But under most circumstances, if you do drink milk, take it alone!

8. EAT FOODS THAT YOU LIKE

Food must be enjoyed in order to promote better enzyme-catalyst digestion. The taste and fragrance of foods, if agreeable, cause the salivary and gastric enzymes to flow. Chewing makes your mouth water and your stomach secrete. This "appetite juice" will boost dynamic enzyme-catalyst regeneration. Foods should be wholesome, of course, but they should be ones that you like.

9. EAT FRUITS EVERY DAY

Fruits are good sources of very vigorous enzymes that can digest tough foods and cause improved distribution of enzymes throughout your body. Plan to use as many fresh fruits every day as is possible.

10. EAT RAW VEGETABLES EVERY DAY

Fresh vegetables are rich in vitamins, minerals, plant proteins and many other essential life-building elements. They are also prime sources of powerful enzyme-catalysts that your body needs for regeneration. Plan to eat raw vegetables every day, preferably with each meal.

11. DRINK RAW JUICES DAILY

They are high concentrates of powerful enzymes that are quickly absorbed and assimilated by your body. Raw fruit and vegetable juices offer you a speedy and almost "instant" source of enzymes for cellular regeneration.

12. EAT FRUITS AND VEGETABLES AT SEPARATE INTERVALS

Enzyme combustion appears to be more favorable if these two foods are *not* eaten at the same time. Fruit enzymes are strong. Vegetable enzymes are somewhat weaker. If combined, there is a dilution and subsequent weakening of all enzymes. Therefore, you would do well to plan to eat a raw vegetable salad *before* your meal; then eat a raw fruit salad *after* your meal. The interval between has given your enzymes an opportunity to catalyze the vegetables, and they now are able to catalyze the fruits without any dilution. You will find that your entire body will become more energized and revitalized if these two high-enzyme foods are *not combined* or not eaten too closely together.

13. AVOID ENZYME-DESTROYING CONDIMENTS

These include mustard, pepper, pepper sauce, salt, cayenne, vinegar, hot-irritating sauces of all kinds. All volatile and burning condiments should be eliminated. They cause gastric distress. They "burn" enzymes and cause tissue destruction. Instead, flavor your foods with wholesome herbs for a delectable and soothing taste.

14. AVOID EXCESSIVE USE OF PROCESSED FOODS

Whether dehydrated, canned, frozen, or otherwise processed, these foods contain almost no enzymes. The very act of processing (high heat and chemicalization) has destroyed these precious catalysts. While it may be difficult for you to completely eliminate packaged foods from your enzyme-catalyst health program, it would be wise to use them at a minimum.

15. AVOID SUGAR

Not only is this refined carbohydrate totally useless (it offers calories and nothing else), but it tends to cause digestive upheaval. It also interferes with proper enzyme-catalyst function. If you must satisfy your sweet urge, do it with luscious naturally sweet fruits and their juices. A bit of honey or molasses (pure) or maple syrup (pure) or a sweet-tasting herb or spice (cinnamon, sweet ginger) would do well. Readjust your eating plans. Retrain your taste buds to enjoy the natural flavors of foods and beverages without sugar in any form. You'll soon be able to boost your enzyme-catalyst power and general health on this sugar-free program.

16. DRINK SIX GLASSES OF BEVERAGES DAILY

You may want to drink water. You may also enjoy fresh fruit juices and vegetable juices. Plan to take in at least one to two quarts of liquids daily. Do this *between* meals. You'll provide needed moisture for your thirsty enzymes. You'll also lubricate your organs and enjoy a healthier and toxemia-free body.

17. BOOST DAILY INTAKE OF FIBER

Known as cellulose or roughage, fiber is important as a "broom" to sweep out wastes from your system. Whole grain wheat germ or bran, taken with a fruit salad, or blenderized in a fruit or vegetable juice, will give you important bulk. You'll be improving your enzyme-catalyst action directly upon your colon to create healthy regularity. You may also sprinkle these grains on a fruit or vegetable salad; the cellulose of the plant foods will add more needed bulk to your system, along with the roughage-forming grains. Enzymes will then be able to help you establish regularity.

18. KEEP YOURSELF WELL VENTILATED

Take frequent "air baths." That is, breathe deeply throughout the day or do simple deep breathing exercise. Remember that your chest is a blood pump as well as an air pump. This deep breathing will "ventilate" or "oxygenate" your enzymes so that they can perform their catalyst work more efficiently.

19. EAT AT REGULAR HOURS

This helps maintain normal digestive rhythm. Your body clocks are able to function better if you establish regular eating times. If you feel like omitting a meal, then eat a platter of fresh raw fruits or vegetables; this maintains your biological rhythm, introduces important enzyme-catalysts, and creates "enzyme harmony" for improved health.

20. HEALTHY SLEEP = HEALTHY ENZYMES

When you refresh your body with healthy sleep, all of your body processes respond with renewed vigor. Plan to have at least eight hours of sleep each night. When you do, enzymes are able to replace cells with new ones much faster than during waking hours. Enzymes will also help promote a cleansing of waste products, and formation of important collagen. Ever notice how haggard sleep-deprived people look? It indicates a deficiency of collagen which forms more rapidly during sleep. So a good night's sleep is as important as good food in order to boost your enzyme-catalyst functions.

Feels "Forever Young" in Eleven Days

Everyone told dress designer Agnes J. that she was looking tired. Someone caustically remarked that her designs looked younger than she did! Agnes J. felt distraught. Crease lines marred a once-smooth skin. Bags started to form under her eyes. She walked too slowly. Her hands trembled. She had to squint to see fine print. A needle and thread became tiresome for her hands! She had to admit that she was aging. What could she do? One of her customers who had been to an expensive and exclusive health spa, suggested she follow the simple enzyme-catalyst program. The preceding 20 principles were used as a basic guideline. Agnes J. made the simple adjustment in her daily eating and living program. *Works Almost Immediately.* The increase in raw foods, the intake of healthier cooked foods and better living habits started to improve her health from the first day. Her enzyme-catalyst system supercharged her organism with dynamic vitality. Her skin firmed up. The bulges under her eyes firmed up, too. She walked more actively. Her hands were strong. Her eyesight so improved, she could see distances without any strain and could even put aside her eyeglasses! By the end of just eleven days, she felt and looked "forever young." Now everyone raved about how she could be her own "younger sister"! All this because enzyme-catalysts had brought about the rejuvenation of her entire system.

HOW TO DISCOVER YOUR "LOST YOUTH" ON THE ENZYME-CATALYST PROGRAM

What causes aging? The absence of important life-bestowing enzymes is the basic cause. But what does this mean in terms of health? Actually, one vital purpose of enzyme-catalysts is to guard against excessive or prolonged toxemia and enervation. These may well be considered the underlying causes of aging. How can you protect against these erosive destructive reactions? By following the enzyme-catalyst program as outlined in this chapter and throughout the book. In so doing, you help protect against such health-destroying factors.

Cause Of Toxins. An accumulation of toxins will cause toxemia, or an accumulation of poisonous substances in the body. To begin, toxins accumulate almost from infancy. During the process of maturity, toxins tend to remain in the body. In the nervous, tense, highstrung person, more toxins tend to gather. They are considered the byproducts of tissue wear. Your entire body is ceaselessly undergoing changes during every moment of life. Old cells wear out, become useless and are replaced by new cells. This is the basis life process.

Problem: The waste resulting from ordinary (even unnatural) wear and tear in these cells does not always find its way out of your body. Much may remain within your organism. The waste products cling to your organs, and in varying amounts, to your trillions of cells and tissues. As a consequence, these molecules become "festers" that impinge and restrict free circulation and improved distribution of nutrients. They *block* total passage of these oxygen-borne nutrients. This leads to the problem of *toxemia* and *enervation.* This means there is a loss of energy as cells, tissues, organs and systems begin to grow sluggish and weak. This sets off your so-called aging process.

Toxins Are Everywhere. Intestinal toxemia is the basic problem. But toxins are spread all over. Some are stored in your glands; more become stored in your cellular spaces, circulating in your blood and lymph, clinging to your vital organs. This blockage takes its toll; as it slows down or otherwise interferes with full metabolism, your body begins to deteriorate. Gradually, your vital organs begin to weaken. Your senses become dulled. Your energy drains away. This is the onset of premature aging. As more and more toxins accumulate, the entire body becomes sluggish and "old." This is the inevitable consequence of accumulated wastes that destroy your basic health. It need not be this way. It can be halted, corrected and eventually eliminated with the use of regular enzyme-catalyst programs.

ONE-DAY-A-MONTH ENZYME-CATALYST FAST
FOR TOTAL REJUVENATION

Basic Program: Just one day a month on a simple enzyme-catalyst fast will work miracles in helping to scrub away the draining toxins, and help promote internal cleansing, the key to total rejuvenation.

Basic Schedule: For this one day, plan to have a variety of seasonal fresh raw fruits and vegetables, as well as their juices. Then follow this easy schedule:

Upon Awakening. Have a glass of any desired citrus juice or fruit juice combo.

Breakfast: A platter of fresh fruits, according to taste and personal desires.

Midday: Two glasses of any fresh vegetable juice or desired combo.

Lunch: A raw vegetable salad. Several raw vegetables such as shredded carrots, green peppers, sliced cauliflower, celery 1chunks.

Midafternoon: Two glasses of any fresh fruit juice or desired combo.

Dinner: A platter of raw vegetables that require a lot of careful and delicious chewing.

Late Evening: A glass of fresh vegetable juice.

Benefits: Just one day per month on this program will help loosen, dislodge, dissolve and eliminate many of the accumulated toxins that are cluttering up your trillions of body cells and tissues. The fruit juices will have a more potent effect on scouring your body cells. The vegetable juices will then liquefy these wastes and distribute them via your eliminative channels for expulsion from your body. The chewy good fruits and vegetables will accelerate the release of more salivary and digestive enzymes that will further improve the metabolic and assimilative powers of your system, and your basic processes will be exhilarated. You will emerge with the look and feel of refreshment. There will be a feeling of sparkling cleanliness in your body that will radiate in the very glow of life as you meet the responsibilities of the days ahead with renewed vigor.

Boost Raw Food Intake for Lasting Health. In addition to this one-day-a-month enzyme-catalyst fast, you should give your body the working materials required for keeping you healthy throughout the rest of the month. Plan to boost your raw food intake throughout your eating plan. Each day, some raw fruit, some raw vegetables, seeds, nuts, whole grains and, of course, fresh juices, will provide you with a treasure of enzymes that are available *on a regular basis* for this scrubbing away of wastes. It is the healthy and tasty way to boost your powers of lasting health.

Overcomes Allergies, Triples Energy, Glows Youthfully

Lila McD. was a sufferer of recurring allergic attacks. She was often so drained that she could scarcely get up out of bed to perform usual household tasks. She looked aged. Lila McD. had tried various prescribed and patent medicines. They offered some easing of symptoms, but made her so groggy, all she wanted to do was sleep. When the medications wore off, her symptoms returned with worsening severity. Lila McD. felt she was doomed to this constant drain upon her energies. She told a visiting health nurse that she felt like an invalid! The nurse suggested she try a simple enzyme program. Go on a raw food fast one day a week. As health improves, if desired, limit it to one day a month. Meantime, boost intake of raw foods throughout the month. Desperate, Lila McD. followed the program. In ten days, she was no longer allergic. In twenty days, she had triple energy. At the end of the month, she glowed with youthful health. Now, Lila McD. enjoys a raw food fast at least twice a month. She feels that life is definitely well worth living since she has freed herself from her weakening problems. Enzymes had boosted body health to a youthful level again.

With the use of enzyme-catalyst foods and simple improvements in your daily lifestyle, you can supercharge your body with dynamic power and vigor.

MAIN POINTS:

1. Follow the 20 step plan to a longer and healthier lifestyle with the enzyme-catalyst program for daily use.
2. Agnes J. became "forever young" in eleven days on a simple program.
3. Lila McD. discovered freedom from allergies, triple energy and a new youthful appearance on a simple enzyme program.

CHAPTER 17

A Treasury of Enzyme Healers for

Common and Uncommon

Everyday Health Problems

The cell-scrubbing power of enzymes will work internally as well as externally. You have a harvest of fresh foods available as natural healers. They may even be comparable to natural medicines because they stimulate your body's sluggish responses into activated metabolism and more youthful vitality. Here is a collection of enzyme healers that will help make you over, outside and inside, and enable you to enjoy the best that life has yet to offer.

Oily Hair. Beat two egg whites until stiff and then apply to your scalp with an old toothbrush. Let the eggs dry and then brush thoroughly before shampooing.

Dry Hair. Rub castor oil (or any vegetable oil) into your scalp at bedtime. Shampoo in the morning. (Reserve an old pillow-case for this healer.) Treat twice a week for several weeks, then once every two weeks. It will make your hair shiny and healthier. *Note:* If you'd prefer not to leave the oil on all night, rub it well into your scalp and steam it in by pressing a hot towel on your head. Shampoo well.

Dandruff. Try mixing equal amounts of vinegar and water, part your hair and apply to your scalp with cotton before shampooing. An alternative method to be used weekly: Beat one raw egg lightly with a fork and rub it well into your scalp, using it as an enzyme-catalyst substitute for shampoo. Then rinse well with warm water.

Face Cleanser. Egg white is a high enzyme-catalyst for a basic facial and skin toner. Just clean your face well and rub with egg white. It will tighten on your skin and require plenty of rinsing, but it leaves your complexion rosy, glowing, immaculate.

Paste Masque. Excellent for tightening up those enlarged pores. Make a paste of raw oatmeal or cornmeal by mixing with a little water. Spread on your face and let remain for at least 15 minutes. Rinse with warm water and then cool water.

Oily Skin. Avocado blended to a paste is a great enzyme-catalyst for scrubbing away clogged pores and soothing oily skin.

Tired Eyes. Soak gauze in freshly squeezed orange juice and apply to your lids. Keep the gauze wet with juice and let it remain on your lids about thirty minutes.

Skin Pep-Up. Cucumber enzymes are used as an ingredient in many expensive modern creams. You can use them for a skin pep-up with an ordinary cucumber. Rub cucumber slices all over your skin. Or, rinse your face with cool water in which you've mashed some cut-up cucumber.

Dry Skin. Rub your face lightly with any cold-pressed oil. And…the juice of a honeydew melon is a great enzyme-catalyst remedy for dry skin; just mash and use as a mask. Or mix the mashed melon with some cold-pressed oil and spread on your face.

Hand Care. Refresh your hands with a rubdown. Just use a lemon wedge or a slice of raw potato.

Chapped Hands. Try oatmeal paste. Wet raw oatmeal to a paste consistency, rub on your hands, leave for a few minutes and wash off. The enzyme-catalyst action leaves them wonderfully soft.

Elbows Rough Or Red. Just rest each elbow in half a grapefruit while you are reading.

Enzyme-Catalyst Bath. Mix one cup of cold-pressed oil with one tablespoon of herbal shampoo. Add several drops of perfume if you want a scented mixture…or you might add oil of rose geranium, lemon verbena or any other scented oil. Blend for a few moments on the highest speed of your blender; or beat with an egg-beater until well blended. Bottle the oil and use four tablespoons to a full tub of water. Enzymes will catalyze away wastes from your skin; the warmth will open your pores so enzyme catalysts can actually steam out wastes from within. You'll feel remarkably refreshed in just thirty minutes.

Diet Aid. Enzyme-catalyst vegetables are invaluable diet aids; carrot sticks, cucumber sticks, radishes, celery sticks. So is fresh grapefruit. The sections are wonderfully juicy and refreshing; the enzymes are able to dissolve accumulated fats and promote better metabolism. *Tip:* Have a bowl of citrus fruit sections handy and nibble a few between meals to reduce the temptation of high-calorie snack attacks.

Appetite Control. Drink a glass of fresh citrus juice before your meal or whenever the urge to eat strikes. The enzymes will help raise your blood sugar and you'll feel satisfied...even without eating. It's the natural appetite suppressant!

For Healthier Hair. A few drops of freshly squeezed lemon juice added to your favorite herbal shampoo leaves your tresses shinier and squeaky-clean. And fresh lemon rinse helps cut dulling soap film from your hair as the enzymes bring out those hidden shiny highlights. (Especially good for oily hair.) Just squeeze the high enzyme juice from half a lemon, strain, then add to a glass of warm water. Pour directly on your hair and massage through. Then rinse thoroughly with clear, lukewarm water.

Enzyme-Glowing Face. Lemon enzymes are a natural astringent. They help tighten pores and protect against excessive oiliness. After washing your face, a little fresh lemon squeezed and strained in your rinse water will help get rid of every trace of soap. Then rinse over and over with clear water.

Enzyme-Softened Hands. Keep a lemon half near the sink to rub over your hands after washing and also to help remove stains and food odors.

Smoother Feet. A wedge of fresh lemon rubbed on the skin will give you refreshing, tingly relief from tired feet. Or soak your feet in a mixture of hot water and the juice of one whole lemon. Rinse in warm water and dry carefully. The enzymes will help whiten and soften your skin while it soothes.

All-Over Feeling. To feel clean, smell clean and tingle all over, drop fresh lemon slices into your bath water. It's like taking an enzyme-catalyst bath! The fragrance will be as delightful as a lemon orchard as your body basks in enzyme-catalyzing lemon luxury. You'll feel renewed all over.

Enzyme Facial Mask. About once a week your face needs a special scrubbing to get rid of the wastes that cause a dull, muddy look. You need an enzyme facial mask. Here's one that really rejuvenates your skin and helps it feel youthfully fresh and cool. Mix together the juice of 1 lemon and the white of 1 egg. Add dry oatmeal gradually until you have a soft paste. Mix with a slight chopping motion, then allow to set a few minutes till the moisture is absorbed. Apply to your face, avoiding areas around your eyes. Let dry about ten minutes, then rinse off with clear, warm water.

Enzyme Refresher. To give your face a little pick-me-up on a hot day (or any day), fill an ice cube tray with equal parts strained lemon juice and water. Freeze. Lightly rub one of the lemon cubes over your face and neck. Rinse with cool water and pat dry. Leaves your face feeling as refreshed as if you'd splashed it in a mountain stream.

Enzyme Tonic. A good thing to keep in your refrigerator at all times is a bottle of an Enzyme Tonic. (For outside, not necessarily inside use.) Mix it up in any quantity you like, using this formula: for every cup of strained lemon juice, add half a cup of water. Then use it as a base for other enzyme formulae. Saves you the bother of squeezing a lemon every time. Keeps well when refrigerated.

Hypertension. Try fresh fruit. Enzymes use its potassium to help soothe your nervous system and ease the risk of high blood pressure. You'll help protect yourself against hypertension more effectively than merely cutting back on salt.

Constipation. Figs! Enzymes use the seeds of the ripe dried fruit to create a gentle intestinal stimulus to encourage natural regularity. *Suggestion:* After breakfast each morning, have several figs followed by a glsss of orange juice. You'll be introducing a rich concentration of enzymes that will promote regular elimination.

Speedy Energy. Again, turn to the enzyme-packed fig. During the ripening process, the natural sugar content of the fig increases quickly. Simple sugars or monosaccharides, in the form of dextrose and fructose, are simple carbohydrates that require no digestion. Your enzymes will cause them to be absorbed instantly into your bloodstream, providing almost instant energy without the fermentation or distress that so often follows eating refined sugary foods that are also enzyme-deficient.

Yogurt Face Mask. Coat your skin with a generous amount of plain yogurt and rub it in gently. If possible, let the enzyme-rich yogurt remain on your face overnight. When you awaken, your skin will glow with youthful freshness.

Basic Enzyme Elixir. Into one glass of orange juice, stir one teaspoon of brewer's yeast and one teaspoon of lecithin granules (from health stores or pharmacies). Stir or blenderize. Drink slowly. You'll have a great supply of energy-boosting and nerve-soothing enzymes along with B-complex vitamins and also Vitamin C. Enzymes will use these nutrients to help create collagen, the plumping layer of your skin. Enzymes will also use collagen, made from the nutrients in this *Elixir,* to build up the connective tissue that holds your body cells and tissues together and protect against the appearance of wrinkles and creases. In a matter of moments, the enzymes will help you feel pepped up, energetic and youthful. You'll look it, too!

Quick Energizer. For a powerhouse of energizing enzymes and fructose, combine plums, cantaloupe balls, watermelon wedges and grapes. Sprinkle with lemon juice to guard against darkening. (Also provides concentrated enzymes and Vitamin C.) Eat slowly with a spoon. Quenches your thirst and peps you up, too.

Quit Smoking Tip. Soothe your frazzled tobacco-tortured taste buds with this milk shake. It's almost fat-free. Combine one glass of skim milk with one banana and one tablespoon of brewer's yeast powder. Blenderize for one minute. Sip slowly. The banana enzymes use the milk minerals and yeast B-complex vitamins to soothe your taste buds, ease your nerves and give you release from the urge to smoke. Try this milk shake twice or three times daily. It enzyme-nourishes your body and builds better resistance to enzyme-destroying tobacco!

Diarrhea. To a glass of any desired fruit juice, add one-half teaspoon of honey and a pinch of kelp or sea salt. Take one glass of this combo after each trip to the bathroom so that the fluid intake is equal to or greater than the amount of water lost with diarrhea. The enzymes will use the fructose and the Vitamin C to help nourish and repair the broken collagen that may be leading to a condition of looseness. Just one day with this program should help control the problem.

Lack of Energy. Here is a homemade health potion that is power-packed with enzymes to give you that "get up and go" feeling. Blend or stir vigorously together: 1/2 cup grape juice, 1/2 cup orange juice, 1 teaspoon lecithin granules, 1 teaspoon brewer's yeast. Drink early in the morning for a burst of energy that lasts and lasts and lasts. It works because enzymes in the grape juice use its high concentration of energy-boosting fructose to supercharge your body with speedy vitality.

Boosting Enzyme Intake. Sprinkle sunflower seeds on your salad. Try sesame seeds on buttered whole wheat bread. Chop or coarsely grind several almonds and sprinkle them over a dish of hot cereal. Mix ground almonds with raisins for a terrific enzyme boost. Add a teaspoonful of wheat germ oil to your favorite salad dressing. Mix some wheat germ into whole grain cereals and baked goods. Sprinkle chopped seeds or nuts and/or raisins onto salads and in whole grain cereals. You'll be enjoying great taste and a super-charging of enzyme-catalyst healing and revitalization from head to toe.

Sensitive Feet. Soothe the soles of your feet (if they look and/or feel red and irritated), with this enzyme-catalyst remedy. Add three teaspoons of lemon or lime juice to one quart of water. Rinse your feet in this solution twice a day. It offers you soothing refreshment. When your feet feel better, so do you...all over!

Skin Pick-Up. Rub a slice of a chilled tomato over your entire face, throat included. Let juice (prime source of skin-cleansing enzymes) dry. Rinse off with tepid water.

Puffy Face. Beat egg white until stiff. Now coat over entire face, from hairline to base of throat. Let dry about ten minutes so enzymes can scour away debris and wastes. Rinse off with cool water.

Enzyme Face-Lift. Mix equal amounts of ordinary honey and egg white. Spread over face from hairline right down to your throatline. Let remain ten minutes. Rinse off with warm water. Finish with a cold water splash. Helps your face look "lifted."

Control Drinking Urge. Satisfy your taste buds with a tangy beverage: to a glass of tomato juice, add juice of one lemon. Add ice cubes, too, if you need this taste feeling. Stir vigorously. Sip as slowly as you would a drink. Repeat often and you'll help ease and erase the urge to drink.

HIGH-ENZYME HERBS INSTEAD OF SALT FOR IMPROVED HEALTH

As fresh raw foods, herbs (regardless of size) do pack a punch in terms of enzyme-catalyst reaction. They can cause total rebuilding to happen in a little while. You would do well to use herbs as flavorings instead of salt which acts as a corrosive and not only "burns" tissues and destroys enzymes, but is responsible for the so-called aging process. So look to herbs as (1) enzyme-catalyst source and (2) healthful taste-satisfying salt substitute.

Basic Suggestion: Rely less on canned, "convenience," highly processed and restaurant foods and more on home-prepared meals made from scratch. You'll have more control over your salt intake. You'll also be able to boost your herb and spice intake. These tiny flavorings are great tasting and even greater sources of enzyme-catalysts. They enable you to actually feast your way to improved health.

HERBS AND SPICE AND EVERYTHING NICE

Herbs and spices can be an excellent means for using less salt in your food preparation, while adding appealing new flavors.

Enjoy experimenting with herbs and spices in your cooking. The following tips will get you started:

Using herbs:

Start with 1/4 teaspoon dried or 1 teaspoon fresh herbs for a dish that serves 4 people. Increase the amount, if desired, to suit your taste.

When soups and stews or other dishes are to be cooked for a long time, add herbs during the last hour of cooking.

Fresh herbs are best, and some can be grown indoors all year long. You can also preserve fresh herbs' flavor by freezing them. Simply remove the stems, seal the cleaned, fresh herbs in airtight plastic bags, and freeze. When needed, snip or chop without thawing.

To receive the full benefit of dried herbs' flavor, keep these points in mind:

Crumple dried leaf herbs between your fingers to release the essential oils.

Add dried herbs to cold foods, such as tomato juice, salad dressing and cottage cheese, at least 24 hours before serving.

For foods that require a short cooking period, soak dried herbs in a small amount of the liquid or oil specified in the recipe for about 1/2 hour before use.

Dried herbs tend to lose their flavor, especially if stored near a warm stove. Keep them in a cool, dry cupboard.

The "correct" combination of herbs for any single food is the one that tastes best to you. Here are some "tried and true" herb combinations:

| oregano and marjoram | mint and marjoram | thyme and parsley |
| tarragon and chives | oregano and rosemary | sage, savory and parsley |

Using spices:

Use 1/4 teaspoon spice per pound of fruit, meal, etc., or per pint of sauce, soup, pudding, batter, beverage, etc.

Add *ground* spices:

to short-cooking dishes when the salt (if any) would be added.
to long-cooking dishes at the end of the cooking period.
to cold dishes several hours before serving.

Add *whole* spices to long-cooking dishes at the beginning of the cooking period. Pulverize or crumble to release flavor.

Uses Seasonings to Improve Basic Health

Roy F. loved tangy foods and beverages. He would use excessive amounts of salt before and after their preparation. Furthermore, he was "addicted" to canned foods which had high chemicalized salt content. These salts eroded his cell-tissue network and reduced effectiveness of enzymes (they become burned in the presence of salt) so that he became old before his time. Even his hands started to shake. To avoid further deterioration, Roy F. followed an herb-spice program offered by his nutrition counselor. It called for the elimination of salt in the diet. Foods were to be fresh. Seasonings could be in the form of herbs and spices. Results? In three weeks, Roy F. seemed "in the pink" again. His grip was firm. He felt as if twenty years had been lifted from his life. He had kicked the anti-enzyme salt habit and thanks to high enzyme-catalyst herbs and spices, he was "reborn."

LIVE LONGER, LIVE BETTER — THE ENZYME-CATALYST WAY

It is important to live *longer.* But it is even more important to live *better,* in terms of looking and feeling youthful, enjoying a happy lifestyle that is free from illness. With a cell-tissue system that is regenerated around the clock by enzymes, this can be a reality. These miracle workers of self-rejuvenation can help you enjoy the best that is yet to come.

Begin today to revive your body and mind with the dynamic enzyme-catalyst health program.

WRAP-UP:

1. Use enzymes, inside and outside, to make yourself over...often, in minutes.

2. Switch to flavorful herbs and spices as a salt substitute and boost your enzyme powers.

3. Roy F. used these power-packed enzyme seasonings to improve his basic health, to return to feeling "in the pink."

ENZYME REVITALIZATION INDEX

To find a quick answer to your problem, just let your finger "dial" down the column on your left. Find your problem then "dial" the page and you have the remedy. . . in a matter of seconds.

LOOK FOR YOUR HEALTH PROBLEM   "ECD" SOLUTION SOURCE

A